Grassland
Butter & Cheese
John Wuethrich
CREAMERY CO.
22
GHT
OMS

BIG RIGS of the 1970s

RON ADAMS

MOTORBOOKS

First published in 2007 by Motorbooks, an imprint of MBI Publishing Company, Galtier Plaza, Suite 200, 380 Jackson Street, St. Paul, MN 55101-3885 USA

Library of Congress Cataloging-in-Publication Data

Adams, Ronald G.
Big rigs of the 1970s / Ronald Adams.
p. cm.
Includes index.
ISBN-13: 978-0-7603-2548-3 (softbound)
ISBN-10: 0-7603-2548-0 (softbound)
1. Tractor trailer combinations–United States–History. I. Title.
TL230.5.T73A333 2007
629.224–dc22
2006019834

Editor: Amy Glaser
Designer: Christopher Fayers

Printed in China

About the author:
Ron Adams is a lifelong big truck enthusiast who has spent many years amassing an archive of vintage truck photos. This is his fifth book for Motorbooks, following *100 Years of Semi Trucks*, *Big Rigs of the 1950s*, *Big Rigs of the 1960s*, and *Mack Trucks*. He lives in Lenhartsville, Pennsylvania.

On the cover, clockwise from top:
The west Texas landscape always provides a scenic environment for photography. This mountain setting makes a nice background for an early 1970s White Freightliner tractor and Brown trailer. The Old El Paso brand of fine Mexican foods is proudly displayed on the rig owned by the Hertz Rental Co. and leased by the Mountain Pass Canning Co. of Anthony, Texas. *Old El Paso*

Icelandic brand frozen seafood, known as the world's finest fish, was hauled in style in the Fruehauf reefer trailer pulled by a White Freightliner tractor. The colors of turquoise, blue, and white make for this cool-looking rig owned by Coldwater Seafood Corp. of Cambridge, Maryland. *Harry Patterson*

This Kenworth W-900 is part of the Farnsworth Trucking fleet, based in Orem, Utah. The Trailmobile reefer trailer wears Meadow Gold Dairy Products graphics. Polished aluminum wheels, a lot of chrome, and pinstripes make this rig stand out. *Author collection*

In 1978 Mack introduced the new Superliner that was its competition to the western manufacturers. Not all Superliners went to the West Coast. This one went to the Cannon Muskegon Corp. in Muskegon, Michigan. The big 1,500-square-inch Radiaton gives this tractor a handsome appearance, along with the colorful Trailmobile trailer. The photo was taken at the Trexler 76 truck stop in Kuhnsville, Pennsylvania. *Ron Adams*

In 1976 many companies celebrated the United States' bicentennial, including South Hills Movers Inc. This Mack F-700 and Fruehauf moving van trailer proudly display their patriotic colors. *Author collection*

On the frontispiece:
The White-Freightliner Vanliner came on strong after it made its debut in 1968 and grew stronger over its first four years. It had 18 inches more room than the standard 86-inch cabs. John Wuethrich Creamery Co. of Greenwood, Wisconsin, hauled Grassland butter and cheese in the Timpte reefer trailer. *Harry Patterson*

On the back cover, clockwise:
The White Western-Star was only seven years old in 1974. It had come a long way in such a short time. The tractor is leased to Delta Truck Lines Ltd. of Toronto, Ontario, Canada. Most of the time you can tell the Canadian tractors by the spread of the tandem. This one has the 120-inch BBC. Notice the extra-long fuel tank. The paint scheme is one of White Western Stars' own designs. *Harry Patterson*

This Ford 8000 was busy delivering fuel for the Union 76 Oil Co. The standard engine was the 175-V8 hooked with a five-speed transmission. The front axle was 9,000 pounds and the rear axle was 32,000 pounds. The wheels were the standard cast spoke. The cab on this truck is a 105.3 BBC. Wheelbases went up to 236 inches. *Author collection*

There were many different moving companies. Not every one was as big as Allied, Mayflower, Bekins, North American, or Global. One of the smaller companies was Cartwright Van Lines Inc., which was based in Kansas City, Missouri. Dromedary rigs were popular with the moving companies at the time. K-100 Kenworths were the power source with many moving companies. The Dromedary body and trailer are of unknown brands. *Harry Patterson*

CONTENTS

DEDICATION

Driving a truck is the ambition of many young people, and they usually don't care what kind of a truck it is, as long as it has wheels. This is how it went for a young man named John Hiestand. John started his driving career in October 1959 for L. H. Liles in Lancaster, Pennsylvania. He drove a Diamond T tandem-axle cab-over-engine pulling a reefer trailer. He made trips to the Midwest and hauled meat on the return trip out of Nebraska and Minnesota to Acme Market in Jersey City. Most of his runs were made over the Pennsylvania and New Jersey turnpikes. He left Liles to drive tanker trucks for Herman Ewell, Inc., until 1966. He then started driving for Roadway Express, Inc., out of its Harrisburg, Pennsylvania, terminal. I got to know John on a Sunday afternoon in the summer of 1968. He was driving a GMC on his way to Worcester, Massachusetts. John drove for Roadway until 1973, when it moved the Break Bulk Terminal to Tannersville, Pennsylvania. John had the opportunity to move with the company but didn't want to relocate his family so he left Roadway and purchased his own truck, a K-100 Kenworth and a flatbed trailer with a side kit. He drove with Interstate Truck Service and worked out of its Reading, Pennsylvania, terminal. In the fall of 1973 he left Interstate and worked in Mason-Dixon Lines' steel division until 1985 and then went on to the company's tank division. John left Mason-Dixon to work for Roadway Package Service. He remained there until the company was acquired by FedEx, where he worked until his retirement in 2004. He spent almost 45 years of his life behind the wheel of a truck and enjoyed his job. I'm sure it wasn't all easy, but, knowing John, he made the best of things and did his job with top honors.

John has always been a truck enthusiast, and still spends time on his hobby. He is president of the Susquehanna chapter of the American Truck Historical Society. To a retired professional truck driver and one of the greatest guys I know, I dedicate this book to John Hiestand.

These W-900 Kenworths became known as legends of the road. They were owned by the Monfort Packing Co. of Greeley, Colorado. The 425-horsepower Caterpillar diesel engines took them down the highways, and most of the time they passed every truck they encountered. These trucks were about the most powerful on the road at the time, and their eye-catching color scheme opened eyes everywhere they pulled Timpte reefer trailers. *Harry Patterson*

CHAPTER 1

1970–1973

A lot of changes in truck models in the 1960s evolved into the trucks of the 1970s. Mack introduced the new F model in 1962. The R model came onto the scene in 1966, including the R and F western models.

White's 4000 and 9000 models were restyled. The new 7000 cab-over-engine model was added to the line around the mid-1960s. International had a lineup of the Stars in 1965. The new CO-4000 cab-over-engine was introduced and replaced the popular Emeryville. In 1967, the Fleetstar's appearance changed. In 1968, the new CO-4070A Transtar was introduced, which was the result of beefing up the CO-4000. The Turbo Star also made its appearance, and in 1969, the Unistar entered the market.

As for Kenworth, the K-100 and the new wide-front W-900 made their debuts in 1963–1964. In 1964, White-Freightliner presented the new half-cab model for the construction industry. In 1968, its new 104-inch cab, known as the Vanliner, was introduced. Peterbilt gave its model 352 a new look; then in 1968, the new model 359 made its debut with a big wide front radiator. Autocar came out with the new S model and the new A64F with the big powerful V-12 Detroit diesel. Brockway also made its contribution with its new 300-series conventionals, including the 359, 360, and 361.

Then Chevrolet brought its Titan 90 and the 70, 80, and 90 series onto the scene. The 90 series had long-

This Autocar, model A64B, was on display at the 1970 American Trucking Association Convention in New York City. Its power came from a Cummins V-903 diesel engine joined with a Fuller nine-speed transmission. It had internal air conditioning. There was a 1,440-square-inch radiator for the big engine, as well as 1,200- and 1,300-square-inch radiators for the smaller engines. The truck's total weight was 12,400 pounds. The photo was taken at Liberty Bell Park in Philadelphia. *Autocar Trucks*

Stars and stripes decorate this R-700 model Mack western series. The power options were Mack, Cummins, Caterpillar, and Detroit diesels with horsepower ranging from 225 up to 475 from the 12V71 engine, one of 26 different engines that were offered. There was a choice of a Mack, Fuller, or Spicer transmission. The tractor and the Fruehauf trailer team up very nicely. *Mack Trucks*

and short-nosed models. Diamond T and REO became Diamond REO in 1967, and its new model was the C-114 conventional. The cab-over-engine models were the CO-78 and the CO-50. The two numbers designated the size of the cab, which was either a 78-inch sleeper cab or a 50-inch day cab.

Dodge came out with its first high cab-over-engine with the 900 and 1000 series in 1965. Ford released the all-new W-series cab-over-engine-style cab in 1966. The year 1969 first saw the new L series, known as the Louisville model, to replace the previous F series. The C series low cab-over-engine continued through the 1960s.

Soon thereafter, GMC brought its new 9500 long-nose and short-nose conventionals into the lineup. Another new tractor from GMC, the Astro 95, was introduced in 1969 with either Detroit or Cummins diesels. Around 1964, Marmon hit the scene with its conventional, and not much later Marmon introduced a cab-over-engine. The White Western Star, which was a custom version of the White 4000 with a few changes, also hit the roads in 1968.

All of these trucks got their start in the 1960s, before many new developments hit the scene in the 1970s, which we'll move on to now.

At the start of the 1970s Autocar offered the A64B and A64F. The A64B was the axle-back version and the A64F was the axle-forward version. The letter *B* designates axle back and the letter *F* designates axle forward. The A75, the axle-back version, was used into the 1970s.

Brockway, which claims to manufacture the most rugged trucks in the world, offered the 300-series 90-inch BBC (bumper to back of cab) short-nose models 358 and 359. Model 360 is the axle-forward long-nose version, and model 361 is the axle-back long nose. Both the 360 and 361 had a 117-inch BBC. The 400 cab-over-engine series included models 456, 457, and 459. All three models were available with the 50-inch BBC day cab or the 90-inch BBC sleeper cab. Model 527, known as the Huskiteer, was Brockway's low-mount cab-over-engine.

Chevrolet produced basically the same models as GMC. The 50 and 60 series were low-tilt cab-over-engine models. The 40, 50, and 60 series conventionals were lighter-duty models, and the 70, 80, and 90 series were heavy-duty models. The first high cab-over-engine for Chevrolet was the Titan 90. The low-tilt cab-over-engine model had a 72-inch cab. The 40, 50, and 60 series used the 96-inch BBC. The 70, 80, and 90 series had a 93-inch BBC short nose and a 115-inch BBC long-nose model. The Titan 90 had a 54-inch day cab, a 74-inch sleeper cab, or an 86-inch sleeper cab. All of

This model C-114 was new from Diamond REO in 1970. Cummins, Caterpillar, or Detroit diesel engines with horsepower from 180 up to 434 were offered, as well as a Diamond REO V-8 250 horsepower and a Gold Comet six-cylinder gasoline engine from 145 to 200 horsepower. Allison, Clark, Fuller, and Spicer powertrains were offered. The steel construction fenders were optional. The last year the big "Diamond REO" name was used on the side of the hood was 1970. *Diamond REO*

The GMC Astro 95 was a somewhat popular truck in the cab-over-engine group. There were nine different engines offered with horsepower ranging from 218 to 335 horsepower in either a Cummins or Detroit diesel. The three cab sizes offered were a 54-inch day cab or a 74- or 86-inch sleeper cab. The model seen here has an 86-inch cab and the standard cast spoke wheels. Red Engine Snacks was the food division of McCleary Industries, Inc., of South Beloit, Illinois. Since snack foods are light in weight, the company decided in the early 1960s that single-axle Fruehauf trailers could handle the job. *Author collection*

There were many livestock rigs on the highways in the early 1970s. This 359 Peterbilt and American livestock trailer was based in Goodland, Kansas. The tractor is wearing the usual owner-operator equipment that was common at the time. The photo was taken at the Union 76 truck stop at Bartonsville, Pennsylvania. *Harry Patterson*

these models kept the same basic appearance through most of the 1970s.

Diamond REO started the 1970s with the C-114 conventional and the CO-50 and CO-80 cab-over-engine. Next in line was the Apollo 116, which had the 116-inch BBC. This was available in both axle-back and axle-forward models. Next came the new Royale, which offered a much larger grille than the previous cab-over-engine. The Royale cab offered a 54-inch BBC cab or an 88-inch BBC sleeper cab. The final model was the Diamond REO Raider, which was offered in two versions—axle back and axle forward, both with a 119-inch BBC. The huge, commanding grille was made of polished aluminum.

Dodge didn't have any new model changes in the first couple years of the 1970s; there were just a few external cosmetic changes However, in 1973, Dodge introduced its new CN-950, known as the Bighorn. It had a 118-inch BBC with a minimum wheelbase of 156 inches.

Ford kept the C-series low cab-over-engine going all through the 1970s. In 1970, the L series, known as the Louisville, was a brand-new model that Ford produced through the 1970s. The LTL-9000 was available with a 118.3-inch BBC or a 154.3-inch BBC with the add-on sleeper box. The LTL was like an owner-operator's truck with many custom options. The W-1000 was manufactured from 1966 to 1970. In 1970, minor cosmetic changes were made and the model number was changed to W-9000. In 1973, the second-generation W series was introduced: the W-900, which had an all-new look with a custom-paint package. This W series lasted until 1977.

White-Freightliners didn't change much in appearance in the early 1970s. In 1973, the new Powerliner debuted, and around the same time, White-Freightliner introduced a new conventional tractor, its first conventional since the late 1940s.

GMC entered the 1970s by continuing the 9500 series that was launched in 1965, along with the 7500 series, a low cab-over-engine model. The Astro 95 was in its infancy but enjoyed steady growth and popularity throughout the 1970s. The 9500-series short nose had a 92.75-inch BBC, while its long-nose big brother had a 114-inch BBC. The 7500 series was the lighter-duty model and also had a 92.75-inch BBC. The Astro 95 had a 54-inch BBC day cab and an 86-inch BBC sleeper cab.

Jones Motor Co., based in Spring City, Pennsylvania, was one of the larger for-hire common carriers. It covered an area from the East Coast to as far west as Missouri. The International COF-4070A Transtar tractors seen here were part of a large fleet. The cabs are 50-inch day cabs. *Jones Motor Co.*

The 5000, 6000, 6500, and 7000 series all had a 97-inch BBC.

International hit the 1970s with a full Stars series: Transtar, Transtar II, Fleetstar, Loadstar, Cargo Star, Paystar, and Unistar. The Transtar had a 51-, 57-, 74-, and 84-inch BBC; the Transtar II had a 59, 74, and 88-inch BBC; the Fleetstar had a 92-inch BBC; and the Loadstar had a 94.5-inch BBC. The Cargo Star was a 70-inch-BBC low cab-over-engine. The Paystar was a 114-inch BBC. The Transtar 4270 and 4370 were a 107- and 117-inch BBC, respectively. The Unistar offered only the 73-inch BBC.

Kenworth entered the 1970s with the K-100 and W-900. The K-100 offered five cab sizes: 52-, 63-, 73-, 76-, and 86-inch BBC. In 1974, a new length, 108 inches, was added. This was the new Vit cab with a much larger bunk. In 1970, Kenworth introduced the new Hustler. This was the low cab-over-engine version that filled many applications in various types of hauling.

At the start of the 1970s, Marmon didn't have a big line of models to offer like the other manufacturers. Marmon produced custom quality-built trucks that offered the same drivetrains and options as other manufacturers. Their lineup included the HDT-BC 86-inch BBC, HDT-AC 86-inch BBC, and CHDT-BC and BCHDT-BC with a 119-inch BBC. Cummins, Caterpillar, and Detroit diesels made up the engine selections, while Fuller and Spicer made up the transmission selections.

Mack carried over many of its 1960s models as it entered the 1970s. The F model offered a 50-, 72-, and 80-inch BBC cab. The Interstater F-series cab had a 50.5- and an 86.5-inch BBC. The U model, which looked a little strange, was the 90-inch BBC offset cab. The U model offered better visibility than most cabs. The R series was available in two versions: the R and the R Western. The DM series included the 400, 600, and 800 models. Both the 400 and 600 had a 98-inch BBC and the 800 had a 122-inch BBC. The MC and the MR were both low-profile cab-over-engine models with setback front axles. The MC had the 59-inch BBC cab and the MR had the 63-inch BBC cab. The MR series was a versatile truck that was designed for refuse, construction, and almost every other application, including oilfield rigging. The MC was used for local pickup and delivery work. Next up is the Cruiseliner, which offered a 54-inch BBC day cab and 76- and 90-inch BBC sleeper cabs.

The C-model Ford cab-over-engine was the longest running model of any manufacturer. This C-model is one of twenty that were ordered, and the trailer is one of fifty new Gindy trailers that were ordered as part of a replacement program for Yale Transport Corp. of New York, New York. *Transport Topics 10-5-70*

Peterbilt brought its two most famous trucks, the restyled 352 and the 359, into the 1970s. Its first axle-back 351 was produced in 1970. It had a 121.25-inch BBC and a Cummins NTC-350 diesel as the standard engine. Model 346 was introduced in 1972 and had a 112.75-inch BBC with a Cummins V-555 as the standard engine. In 1973, the pull-down door handle from below the window was moved to the lower corner on the door as a pull-out handle on the conventional models. In 1972, Peterbilt came out with another first, the new model 352 with the 110-inch cab. Next came the 200 and 300 models. Model 200 was the single-axle version with a 75.5-inch BBC, and model 300 was the tandem-axle version with a 75.5-inch BBC. Both models came with the Detroit 6172 diesel engine. Models 381 and 383 were both off-highway trucks. The 381 had a 121.25-inch BBC and came with the Cummins Super 250 diesel engine, and the 383 had a 139.75-inch BBC and came with the Cummins Super 250 as well. Model 348 dump and mixer trucks both had a 115.25-inch BBC with the Cummins V-555 diesel engine. Model 353 was for mixer, logging, construction, and oil-field work. The BBC on the axle-forward version was 123 inches and the axle back's was 117 inches. The BBC could go to 177.25 inches, depending on the engine selected. The 289 single axle back and the 359 tandem axle back both had a 113.25-inch BBC. Peterbilt came out with 11 new models in the 1970s, and they made only one model of the 248 and 253 in 1976. The model with the highest production was the 352, with almost 29,000 units produced.

White offered the low-profile cab-over-engine Road Xpeditor model, which was an around-town delivery truck. The White Constructor was designed for, as its name implies, construction. Its front axles ranged from 12,000 to 18,000 pounds, while its rear axles ranged from 34,000 to 44,000 pounds. The 4000 and 9000 series both had tilt hoods. The 4000 had a 116-inch BBC and the 9000 had a 106-inch BBC. A wide choice of gas- and diesel-powered engines was available. Another low-profile cab model was the Alley Cat. It was designed for beverage delivery and was a modification of the 1500-series compact. The setback front axle carried 18,000 pounds to give the Alley Cat 6,000 pounds more payload than comparable trucks. A typical beverage truck carries 320 cases of soft drinks, while the Alley Cat carries 450 cases. The Trend is another low-profile cab-over-engine

The K-100 Kenworth was a popular cab-over-engine with many owner-operators. This one is pulling a Great Dane reefer trailer. The rig is leased to Belford Trucking Co. of Miami, Florida, and is hauling fruit and produce to the Northeast and Midwest. Belford Trucking was one of several companies owned by Midwest Emery Freight System, Inc., of Chicago. *Harry Patterson*

from White. The engine is a 327-cubic-inch V-8, and the Trend rides on wheelbases anywhere from 119 to 179 inches. The model 7000 cab-over-engine was offered in a 62- or 86-inch BBC, with wheelbases ranging from 136 to 166 inches. The standard engine was a Cummins NH-180L diesel with a Fuller 10-speed transmission. The new Road Boss was based on the 4000 and 9000 models. It had a standard fiberglass tilt hood and a choice of either 92-, 108-, or 122-inch BBC cabs with engines available up to 350 horsepower. The Road Boss II was a more beefed-up and slightly improved version of the first Road Boss. The Road Commander was the replacement for the 7000 model. It had a 64-inch BBC day cab and an 88-inch BBC sleeper cab. The standard engine was the Cummins Super 250, with nine optional engines available. As the Road Boss II hit the scene so did the Road Commander II. The standard engine was a 300 Cummins with horsepower available up to 435. Both the Road Boss I and II and the Road Commander I and II had a big selection of paint schemes and colors to choose from.

The White Western Star was only three years old as it entered the 1970s, and its great looks didn't change much through the decade. The vehicles worked in all trucking fields and offered the same basic engines and transmissions, wheelbases, custom packages, and options as its competitors did.

Changes in trucking took place in the early 1970s, including the appearance of 42- and 44-foot trailers that eventually replaced the 38- and 40-foot trailers. In the western states 27- and 28-foot doubles were still a big hit with many freight carriers. With the lengthening of the trailers, the truck manufacturers played their part in designing tractors to pull these longer trailers to stay within legal length limits and carry more payload.

The early 1970s also started the age of CB radios and brought on handles such as Strawhat, Bedroom Bandit, Rubber Duck, Smokie, and Leadfoot Lady of Interstate 80, to mention a few. We cannot forget the passing lane that earned the name Hammer Lane and then Monfort Lane, which came from the Monfort of Colorado trucks because of their powerful 425-horsepower Caterpillar engines in those colorful Kenworth trucks. The Monfort trucks were a legend in their time.

DODGE
Dodge Truck
Dodge Truck
1000

THERMO KING
Tropical Plants
and
Plant Care Products
Montreal Canada
4301
WHITE FREIGHTLINER
J.E. Fortin
TRANSPORT

Opposite top: "Wall to Wall Luxury" is what Dodge claims about its cab-over-engine tractor. The cab featured padded door panels, carpeting, easy-to-read gauges, and other driver conveniences. Cummins and Detroit diesels were offered, along with the 413 to 549 V-8 gasoline engines. This one has the 79-inch sleeper cab, and it is also available in a day cab. *Dodge Trucks* **Opposite bottom:** A colorful rig traveling down the highways always catches attention. This rig is no exception. This White Freightliner is owned by J. E. Fortin Transport Ltd. of Canada and is pulling a Great Dane trailer of refrigerated products. The graphics on the trailer tell us that tropical plants and plant-care products were the specialty for Lemoine Tropica, Inc., in Montreal, Canada. *Harry Patterson* **Above:** This RL-700 is in the Mack western class. At the time, there were four engines to choose from: Cummins, Caterpillar, Mack, and Detroit diesel. There were 18 types of engines offered with horsepower ranging from 225 to 475. Phillips Industries, Inc., chose this RL-700 with optional dual stacks and an add-on sleeper box. The trailer is a Fruehauf flatbed with rack sides. *Harry Patterson* **Right:** This dump truck is a Ford Louisville 900. Its power came from a 401 SD V-8 engine and a five-speed transmission. It had a GVW rated at 60,000 pounds. The standard front axle came in at 9,000 pounds, while the rear came in at 32,000 pounds. The GCW was 80,000 pounds. It was a 105.3-inch BBC, and the dump body was made by Gar-Wood. *Ford Motor Co.*

Right: The White Western Star was three years old in 1970. It had the 4800 series and offered the same components as the other truck brands. Sea-Land Services was a container operation for overseas shipping. This White Western Star and Fruehauf container worked on the Pacific coast operation. *Brian Williams* **Below:** Rowley Interstate, Inc., based in Dubuque, Iowa, was one of the many refrigerated carriers that hauled from the Midwest to the East Coast. This model 352 Peterbilt and American reefer trailer was one of the many rigs in the Rowley operation. It was around this time that the drivers started to paint their trailers to match their tractors. The picture was taken around 1970 at the Union 76 truck stop at Bartonsville, Pennsylvania. *Harry Patterson* **Opposite top:** This GMC Astro 95 had a 54-inch day cab on a 142-inch wheelbase. The Brenner tank trailer had a two-door belly-mount cabinet on a two-compartment (2,500 gallons front, 4,000 gallons rear) 6,500-gallon tank body. The trailer was 39 feet long to give the whole rig a 48-foot, 6-inch, overall length. *Brenner Tank Co.* **Opposite bottom:** The amount of milk being hauled in this eight-axle setup is enough to last a family of four for 50 years. The stainless-steel tanks hold 12,500 gallons of milk and weigh in at 107,500 pounds. The tractor is an International CO-F 4090-A Transtar. The power is provided by a 12V-71 Detroit diesel engine with a 13-speed transmission. The rig measures 64 feet, 11 inches, and has a gross weight of 152,280 pounds riding on 42 tires. This rig is one of a pair owned by the Michigan Milk Producers Association. *Transport Topics*

GMC

INTERNATIONAL
TRANSTAR
9459·CK

VERNON
№ 3
Autocar-Diesel
VERNON
RANSPORTATION
BELOIT, OHIO

Opposite top: Marmon trucks, about seven years old in 1970, were somewhat new to the truck manufacturing industry. This model HDT had the 86-inch cab. The standard engine offered was the Cummins NTC-350 with a Fuller RTO 13-speed transmission. The long-stretch livestock trailer is a Wilson. *Wilson Trailer Co. (Woodworth)*

Opposite bottom: This A-75 Autocar of the early 1970s was still going strong when this picture was taken in June 1999 in Emlenton, Pennsylvania. The A-75 had all-aluminum components, and while many manufacturers were going to tilt hoods, Autocar was still using the butterfly hood. This rig is owned by Vernon Transportation Co. of Beloit, Ohio. *John Hiestand* **Top:** Atomic Transfer Ltd., based in Winnipeg, Manitoba, had rigs running throughout Canada. Its fleet was made up of several truck brands: Hayes, Kenworth, International, White Freightliner, and Mack, like this R-600 pulling this converta-flat flatbed trailer. Atomic's colors were metallic green and black. *Ralf Lilje-Gren Photographer* **Above:** The Alley Cat was a modification of the White 1500 compact and was designed for beverage bodies. The setback front axle carries 18,000 pounds, or 6,000 more pounds of payload than comparable trucks. The typical beverage truck carries 320 cases, but the Alley Cat carries 450 cases. The lowered cab makes for easier entrance. The engines offered were the White 6-200A gasoline or the Cummins V-8 185 diesel. Available transmissions were an Allison automatic, a Clark five-speed, and a Fuller six-speed. *White Motor Co.*

Top: From the 1930s through 1970s there was a difference between eastern-type trucks and western-type trucks. One of the differences was that a lot of the western trucks had longer wheelbases than the eastern trucks. This White Freightliner is a good example of a western truck. The owner's name is unknown, but you can see the pride he takes in his equipment. Plenty of chrome and a nice paint job make this tractor noticeable as it goes down the highway pulling the Brown reefer trailer. *Author collection* **Above:** Pitt Truck, Inc., of Augusta, Illinois, chose this Diamond REO C-11464DB to pull the Dorsey reefer trailer. The C-11464 came in two versions: axle-forward and axle-back. This one is the axle-back version, which is what the letter *B* stands for in the model identification. The standard engine was the NH-220 Cummins diesel. Sixteen other engines were available with horsepower ranging from 195 up to 335. Cummins, Detroit, and Caterpillar engines were available. Both Fuller and Spicer transmissions were available with 5 to 16 forward speeds. The longest wheelbase for this model was 224 inches. *Author collection*

Above: When you restore an older truck, you try to picture in your mind what you want the finished project to look like. From the chassis on up nothing goes untouched—from the nuts and bolts to the finished paint job. Jonathan Guss of Salt Lake City, Utah, restored this 1970 model 359 Peterbilt. This truck is typical of the western-type dump rigs and is known as a transfer dump, in which the body on the trailer transfers into the body on the truck when empty. A truck like this is a show stealer. *Don Mackenzie* **Below:** Today's car carrier rigs can haul as many as 11 cars in one load, and depending on the size of the cars, they can carry even more. Due to the size of the cars in 1970, seven was about the limit on the western-type car transporters. The tractor seen here is a Kenworth W-900 model. Notice that the stack is tight against the car over the cab. This 1970 photo was taken in front of the Rose Bowl. *Kenworth Truck Co.*

Right: The C-model Ford low-tilt cab-over-engine was one of the longest-running models of all trucks. This sleeper cab version is a straight truck with an Eager Beaver body hauling new Eager Beaver trailers. The cabs were made by the Budd Co. Mack also used this same cab for its N-series trucks. *Bob Parrish* **Below:** Global Van Lines, Inc., of Anaheim, California, was a national and worldwide moving company. This F-600 Mack and Highway moving-van trailer was one of the many rigs in the fleet. The driver traveled in comfort with an air conditioner and a few fancy extras such as polished aluminum wheels, chrome sleeper vent door, air horns, breather bonnet, and stack. The cab is decorated with all the required state permits. This trailer is equipped with a reefer unit. *Norm Stevenson Photography* **Opposite top:** The International CO-4070 Transtar was about as popular a truck as its forerunner, the Emeryville. This Transtar is owned by Rollins Leasing, as is the Fruehauf flatbed trailer. It is leased by Tull Metals of Atlanta, Georgia, to haul the company's steel products. Several engines and transmissions were available. *Ron Adams collection* **Opposite bottom:** The W-900 Kenworth made its debut in 1956 and is still being made today. Over the years the design was modified. This 1970-era W-900 was dolled up with the options of the time, along with a fancy paint job. The owner of this long-wheelbase tractor and Cornhusker belly-drop grain trailer is unknown. *Neil Sherff*

480
INTERNATIONAL
480
TULL METALS
TULL METALS
Everything In Metals

Above: Hall's Motor Transit Co., Inc., was based in Sunbury, Pennsylvania, and was a for-hire common carrier that covered the Middle Atlantic states and Ohio. One of the trucks in the fleet was this White 9000 and Strick trailer. The photo was taken on October 19, 1970. Here the air shield on the top of the cab shows that fuel-efficiency measures were starting to take place. *Hall's Motor Transit Co.* **Below:** The order of the day was load 'em up and move 'em. Once they were loaded there was no time for fooling around. Midwest Livestock Express of Wilsonville, Nebraska, selected this model 352 Peterbilt and Wilson livestock trailer to get the job done. This company was in its infancy according to the number *1* on the tractor. *Harry Patterson*

Above: The White Freightliner Vanliner was the first of the extended-cab, cab-over-engine tractors. It stretched to 104 inches and was introduced in 1968. Darr Trucking of Fairview, Montana, used this one to pull a three-hopper bottom grain trailer. *Harry Patterson* **Below:** Over the years, many different types of truck bodies were mounted on truck chassis. In this case, we have an F-model Mack western truck with a powerful pumping unit. The unit's main function is oil well stimulation. The truck is equipped with a new model 127-T hydraulic pump unit. Two 100 HPT-902 Saturn gas turbine engines drive the pumps, and each pumps up to 15,000 pounds per square inch. Bryon Jackson, Inc., selected this truck at a cost of $250,000. *Transport Topics*

Above: There were many different moving companies. Not everyone was as big as Allied, Mayflower, Bekins, North American, or Global. One of the smaller companies was Cartwright Van Lines, Inc., which was based in Kansas City, Missouri. Dromedary rigs were popular with the moving companies at the time and K-100 Kenworths were the power source for many of them. The brands of the dromedary body and trailer are unknown. *Harry Patterson*
Right: The White model 7000 was a popular truck with many of the common carriers. Although it is a plain stock tractor, it is sitting on the sales lot of one of the White dealerships. The stock engines were Cummins diesels. *White Trucks*

Above: The W model was a replacement for the previous H model. The W was introduced in 1967. The first four years of the W series had the big Ford letters across the front of the cab. In 1971, Ford changed the design to smaller letters. The W 9000 offered three diesel engines: Cummins, Caterpillar, and Detroit, with 20 different horsepower selections, from 225 to 335. The owner is Kenny Parde, the driver is Mert Parde, and the three companies listed are Freightways, Inc.; East-West Express of Omaha, Nebraska; and Refrigerated Transport, Inc., of Atlanta, Georgia. The dual stacks make me believe that it was powered by a 318-horsepower Detroit diesel. *Harry Patterson* **Below:** The 9500 series came in two versions: long-nose and short-nose. This is the 9500 long-nose. There was plenty of room under that big tilt hood from side to side and front to back to accommodate the big 318-horsepower Detroit and the 335 Cummins diesel. The choices of wheelbases were from 140 to 235 inches. The long-nose was a 114-inch bumper-to-back of cab (BBC). Marvin Martin of Los Angeles, California, had a fleet of these long-nose 9500s. Number 18 has the add-on box sleeper with dual-trailer high stacks. This one is probably powered by a 318 Detroit diesel to pull the Brown reefer trailer. This photo was taken in Arizona. *Brian Williams*

Top: This model F-361 TL Brockway was on showoff time before working time started. There were five different 361 models. The N361 TL was powered by an NH-250 Cummins, the E-361 TL was powered by a 671 N Detroit diesel, the F-361 TL (pictured) was powered by a 1674 Caterpillar diesel, the K-361 TL was powered by the 8V-71 N Detroit diesel, and the V-361 TL was powered by the V-903 Cummins diesel. Each model had the choice of a 5-, 10-, 13-, or 15-speed transmission. The models listed above are the standard engines only. Each model had optional horsepower engines to choose from except the F-361 TL. *Brockway Trucks* **Above:** The Diamond REO is a great-looking truck that worked in all fields of trucking. This model C-116 is known as a transfer dump and is working in the construction field. The body and trailer were made by Truk'Quip in Tulsa, Oklahoma. All the literature and specification sheets show all models with round headlights. The origin of square headlights is unknown. *Author collection*

Left: United Truck Lines and Buckingham Transportation Co. merged and became United-Buckingham, which then merged with Ringsby to become Ringsby-United, based in Denver, Colorado. This White Freightliner and Fruehauf reefer trailer were in the Ringsby-United special-commodities reefer division. Ringsby went out of business in 1976 after a three-year strike. *Ron Adams* **Below:** Earlier we saw a Kenworth dromedary rig for Cartwright Van Lines. Here we have another dromedary rig, but it's a model 352 Peterbilt and Kentucky moving van trailer for Aero Mayflower Transit Co., Inc., based in Indianapolis, Indiana. This rig did a lot of hauling for electronic trade shows. The photo was taken at the Union 76 truck stop in Bartonsville, Pennsylvania. *Harry Patterson*

Above: The Chevrolet Titan 90 was basically the same tractor as the GMC Astro 95. The Titan 90 pictured here is powered by a 12V71 Detroit diesel, which is the largest engine ever used in a Chevrolet. Everything sits on a 150-inch wheelbase. The engine's partner is a Fuller 13-speed transmission. *Chevrolet* **Right:** The A-75 Autocar was the partner to the A-100 and the A-64. The A-75B was the front axle-back version. A few extras and chrome items make this tractor look great. For comfort away from home a sleeper box was added. Produce is probably the commodity hauled in the Great Dane reefer. The owner of the rig is not visible. *Author collection*

The Mack western R series was a popular truck with a lot of western truck operators. The choice of 20 engines ranged from 235 to 375 horsepower from Mack, Caterpillar, Cummins, and Detroit diesels in the R-700 series. Bigheart Crude Oil Corp. of Tulsa, Oklahoma, had this RL-700 to pull the Fruehauf tank trailer. *Bigheart Crude Oil Corp.*

Oshkosh was a longtime maker of four-wheel-drive trucks for military, construction, and snowplowing work. Around the early 1970s, Oshkosh started to make trucks for over-the-road use. Although the sign shows Salt Creek Freightways in Wyoming, it is not known if this model-E tractor is owned by Salt Creek Freightways or if it was just a demonstrator. According to the spec sheet, a Caterpillar diesel was the only available engine. The wheelbase on this model E-1244 was 147 or 165 inches. *Salt Creek Freightways*

The White Motor Co. made construction equipment just like other truck manufacturers. Their offer was known as the Construcktor, with a choice of eight different diesel engines ranging from 195 to 335 horsepower. Transmissions were from 5 to 13 speeds with a 6-speed semiautomatic. The application here was a lift-bed body to haul lumber for Georgia-Pacific in Tigard, Oregon. *Georgia-Pacific Corp.*

FRUEHAUF
Amway
Amway
SHOP WITHOUT
GOING SHOPPING
683
227
227
227
KENWORTH

Opposite top: The DF-8500 International was designed for oilfield use. In 1971 and 1972, only six were built, all at the San Leandro, California, plant. These trucks were powered by a 12V-71 Detroit diesel. A 24,000-pound front axle was an option. The steel front-end components, including the diamond-plate fenders, were available at extra cost. Rear tandems were available up to 90,000 pounds. *International* **Opposite bottom:** Amway products were delivered to the company's warehouse in trucks such as this Kenworth K-100 and Fruehauf trailer. Notice all the license plates that were required at the time for traveling in many different states. This nice, clean rig opened eyes as it traveled down the highway. The truck is owned by Equipment Leasing Co., Inc., of Grand Rapids, Michigan, and is leased by the Amway Corp. *Author collection*

Above: The GMC 9500-series long-nose was a tough-looking truck with the big, broad long-tilt hood. One gasoline and 10 different diesel engines were offered with horsepower ranging from 218 to 335. It had a 114 BBC with wheelbases from 140 to 235 inches. This owner-operator was leased to Caravan Refrigerated Cargo, Inc., of Dallas, Texas. *Ron Adams* **Left:** Our neighbors in Canada also used Ford trucks. Jeffery's Cartage Ltd. of Montreal had this W-9000 model pulling a Can-Car three-axle trailer. Notice the airfoils on each corner of the cab. A total of 13 Cummins, 2 Caterpillar, and 4 Detroit diesels were offered. The sleeper cab was 82 inches BBC. *Author collection* **Below:** This 1971 Dodge LCF 900 went from the old look to the new look through restoration. Three engines were available, and five other higher-horsepower engines were optional. Some extras were added to give it a custom look. The truck is owned by Dusty Memories and Faded Glory of Somerset, Pennsylvania. *Don Mackenzie*

Above: This 1971 series-50 Chevrolet came with the 292 six-cylinder gas engine as standard. A new low-cost Allison four-speed automatic transmission was available for the first time. Wheelbases ranged from 97 to 175 inches, and available engines included one six, three V-8s, two V-6s, and one diesel. The cab was a 72-inch BBC and tilted 55 degrees. It has a 2,000-gallon tank. *Chevrolet* **Below:** The model number is not visible with the door open, but this is probably the Cargostar model COF-1810A. This was the gas model that offered the V-345 at 196 horsepower and V-392 at 236 horsepower. A five-speed transmission was standard with seven five-speed and two six-speed automatics as optional. The International backhoe and bucket loader are being hauled on a Schwartz tilt body. *International* **Opposite top:** The 9500-series short-nose GMC was used by many motor and private carriers. In this case the private carrier is Packaging Corporation of America. The specs on this exact tractor are not known, but in this series there was a choice of nine diesel engines with horsepower from 218 to 335 from either Detroit or Cummins. There was also a 275-horsepower 637 V-8 gas engine and a selection of 10 transmissions. Wheelbases ranged from 140 to 235 inches. This tractor is pulling a Fruehauf trailer. *Author collection* **Opposite bottom:** This Diamond REO may be one of the tractors in the 1970–1971 changeover. Notice that the grille has two chrome pieces down each side and one across the top. This was the style when they were still using the long hood bar with the big Diamond REO letters. The Apollo 116 did away with that and was available in either axle-back or axle-forward. The wheelbases ranged from 158 to 242 inches, with the 242 on the axle-forward model. On the axle-back model, more weight was on the front axle. Engines were Cummins, Caterpillar, and Detroit diesels with 30 models up to 475 horsepower to choose from, and Fuller and Spicer offered nine different transmissions. The tractor and Great Dane reefer trailer were owned by United Produce and Products Equipment Co. based in Birmingham, Alabama. *Brian Williams*

Packaging
Packaging
CORPORATION OF AMERICA
Packaging
Packaging

TRANSICOLD
SPACE MAKER
NO 11
UNITED PRODUCE AND PRODUCTS EQUIPMENT CO.
TENN. HF 7273
U.S.D.A. 60189
105A
104

WHITE FREIGHTLINER
BIG SKY
CO OP
MONTANA
86
28
THERMO KING

J. M. Barrett
LEASED TO
C&H
TRANSPORTATION CO.
DALLAS, TEXAS
MO. M.F.T. PERMIT NO 10256
SD-PUC-7969-A
IDAHO G.W. 74000
600511
FIRE EXTINGUISHER INSIDE

Opposite top: Back in the mid- and late 1960s and early 1970s there was a great number of farm co-ops. One of the more popular was Big Sky Co-Op based in Great Falls, Montana, and Artesia, California, which formed around 1969. The company fleet consisted of Kenworths, White Freightliners, and Peterbilts. Here is a White Freightliner and a Hobbs vanguard reefer trailer. In 1969, the Union Tractor Co. of Dillon, Montana, was also leased to Big Sky. The commodities hauled were farm and nursery products. *Harry Patterson* **Opposite bottom:** C&H Transportation Co. of Dallas, Texas, was a heavy-haul carrier that moved steel castings, machinery, heavy equipment, and other mixed loads. C&H owned some of its equipment, but it had many owner-operators as well. Many loads were hauled in style and were pulled behind this sharp-looking model 352 Peterbilt owned by J. M. Barrett, a lease operator. The power source is probably a 318 Detroit diesel with a Jacobsen engine brake known as the "Jake Brake." *Harry Patterson* **Above:** This 1971 International Unistar was a four-wheel-drive tractor. It was the ideal tractor for pulling doubles in snow, ice, mud, or for field use. Four-wheel-drive operates automatically when 3 percent slippage occurs at the rear-drive axle, which engages the drive mechanism to front-drive axle. The trailers are Titan model 24s. The rig weight is 23,370 pounds, grossing 80,000 pounds, of which 56,630 pounds is payload. This rig was one of Titan's September 1974 show trailers. *Titan Trailers* **Right:** The A64F was Autocar's truck to compete with for western truck manufacturers. It was all aluminum, which reduced its weight. The standard engine was the 12V-71 Detroit diesel that put out 475 horsepower with a 13-speed Fuller transmission. The rear axle at 44,000 pounds and the front axle at 12,000 pounds are Rockwell-Standard. This A64F is leased to Shaffer Trucking of New Kingstown, Pennsylvania, and Lincoln, Nebraska. The sleeper box is a 60-inch Kenworth Aerodyne. The picture was taken in 1988 at the Bandit Truck Stop in New Smithville, Pennsylvania. *Ron Adams*

Left: This White 7000 was busy hauling two new Fruehauf flatbed steel trailers, probably for the Michigan area. Jones Motor Co., Inc., based in Spring City, Pennsylvania, was a rather large common carrier. Notice the short bumper. The reason for this was either to keep it from damaging the tires in case of an accident or to provide more air to cool the tires. *Jones Motor Co.* **Below:** The model 359, introduced in 1967, was as popular as the 351. The 1,444-square-inch radiator could accommodate bigger engines. The owner was W. R. Hensarling Corp. of Madisonville, Texas, which hauled livestock in the Wilson livestock trailer. *Harry Patterson* **Opposite top:** This fleet of LNT-1000 Dodge tractors was kept busy hauling Weavers Country Fried Chicken in Great Dane reefer trailers. Six Cummins diesel six-cylinder engines were offered, as well as two V-8 Detroit diesels. The wheelbases ranged from 122 to 200 inches. The cab was a 79-inch BBC sleeper. A variety of Spicer and Fuller transmissions was available. *Author collection* **Opposite bottom:** By 1971, the International COF-4070A Transtar was selling well. Kaiser Aluminum and Chemical Corp. used one to pull the Trailmobile open-top trailer. The only visible option is the cab-mounted air conditioner. The dual stacks could be either a 6V-71 or an 8V-71 Detroit diesel. *Author collection*

Weaver
Weaver
Country-Style Chicken
DODGE

KAISER
ALUMINUM & CHEMICAL CORPORATION
KAISER
KAISER
INTERNATIONAL
26501

FORD
2219
CENTRAL
3402
J22-496

Opposite top: Central Freight Lines, Inc., of Waco, Texas, was a big Texas common carrier. This 9000-series Ford is hooked to a flatbed trailer and is hauling some kind of pre-cast concrete product. A 52-inch BBC day cab was chosen for this tractor. *Harry Hess* **Above:** Private carriers chose equipment with criteria similar to those used by common carriers. In this case, Thrifty Drug and Discount Stores used the GMC Astro 95 to deliver merchandise to its stores. The engines available were either Cummins or Detroit diesels. The dual stacks suggest that the engine in this tractor is either a 6-71N or an 8V-71N Detroit. Five- to sixteen-speed transmissions were available. This rig was in Thrifty's ice cream division. *Author collection* **Below:** These Diamond REO C-116s were quite an awesome sight. This axle-forward model had wheelbases ranging from 140 to 236 inches in 6-inch increments. Cummins, Caterpillar, and Detroit diesels were available. Here we see a custom package with a sleeper box. All the chrome dresses it up in its Sunday best. Behind the tractor is a shiny new Fruehauf trailer. The rig was based in Utah. *Author collection*

Above: Another Utah-based rig is this White Freightliner for W. S. Hatch Co., of Wood Cross, Utah. This rig is set up as a truck-trailer-tanker combination designed for hauling corrosives. W. S. Hatch covered most of the western states using tankers and dump trailers. White Freightliner trucks and tractors made up most of the Hatch fleet. *Brian Williams* **Below:** Standard Trucking Co. of Charlotte, North Carolina, served both the Carolinas. The company had approximately 250 tractors and 650 trailers working out of 10 terminals. One of the rigs is this White 9000 tractor and Gindy post exterior trailer. The White 9000 was a popular tractor with many fleets. The BBC was 106 inches. Both diesel- and gasoline-powered engines were available. *Standard Trucking Co.* **Opposite top:** This International COF-4070A Transtar was kept busy moving freight in the two Comet trailers in the Garrett system. Garrett Freightlines was based in Pocatello, Idaho, and covered 11 western states plus North Dakota and Minnesota. Garrett pulled double trailers, 42-foot trailers, and triple trailers where permitted. Double trailers were a common sight on western highways. *Garrett Freightlines* **Opposite bottom:** The R-model Macks populated the highways about as much as many other trucks. This one is busy working for the Singleton Packing Co. of Tampa, Florida, pulling a Great Dane reefer trailer. The model number on this one is unknown, but the R-600 had the Mack engine with horsepower ranging from 180 to 270. The R-700 had the choice of Cummins, Detroit, or Mack diesel engines with horsepower from 225 to 335. *Harry Patterson*

GARRETT
42·8774
INTERNATIONAL
42·8774
42·8775

THERMO KING
37
MACK
Mack
94
PACKING CORP

Left: The A-10064 Autocar was all aluminum, which saved weight for more payload. The heavy-duty construction fenders were optional. Of course, the Hobbs tri-axle drop-deck trailer was used to transport construction equipment. *Author collection* **Below:** Parkhill Truck Co. of Oklahoma City was a heavy-haul carrier. The fleet was comprised of owner-operators. All makes of tractors were in the fleet, such as this K-100 Kenworth. Although the headache rack is not visible, we can see some of the chains that were necessary for tying down and securing loads. *Harry Patterson* **Opposite top:** Providing homes with heating oil was the job of the Feece Oil Co. in Batavia, Illinois. The oil was carried in this LN-800 Ford with Progress 2,000-gallon tank body. The L800 had the 361 or the 189 XD V-8 engine with a five-speed transmission. The wheelbases ranged from 138 to 222 inches. The standard wheels on this model were the cast spoke, but here the optional disc wheels were chosen. *Progress Industries* **Opposite bottom:** Standing tall and looking right at you is this GMC Astro 95 and Brown trailer. The rig is leased to the Shatterproof Glass Co. in Detroit, Michigan. The necessary stickers and license plates always decorated many trucks. *Author collection*

1922 J.F.-A.M.
PROGRE
FEECE OIL CO.
FLAMMABLE
A HOME OWNED CO.
38
Phone 879-1911
BATAVIA, ILL.
E.W. 11,500
IMFT 4207

GMC
136
Shat-R-Proof
Shatterproof Glass Corp.
DETROIT, MICH.
2442-CB

STOCKTON, CALIF.
466-2411

Opposite top: Another Diamond REO was this long-wheelbase truck pulling a grain trailer. Notice the 60-inch sleeper box, which was not too common in the early 1970s. The driver has the comfort of an air conditioner. *Harry Patterson* **Opposite bottom:** Peterbilts were not only used as road tractors but were also used for towing. The Connell Motor Truck Co. in Stockton, California, was the owner of this model 359 Peterbilt. Plenty of chrome shows off this great-looking rig. *Author collection* **Above:** A wonderful load of Wonder bread and Hostess cupcakes is ready to be delivered fresh to grocery stores by this White Freightliner with a Utility body. The dual stacks indicate that the engine could be a 318-horsepower Detroit diesel. Notice the stack arrangement. It is not unusual to see the stacks coming up in the position they are in, but what is different is how high the mufflers are and how the mufflers and stacks are mounted to the body, which is rare. To get the most volume for more payload a 48-inch cab was used. Continental Baking Co. owns this breadwinner. *Author collection* **Right:** This company was named after the owner's daughter. Lisa Motor Lines, Inc., of Fort Worth, Texas, was a refrigerated carrier that got its start around the late 1960s and is still running today. The fleet was mainly made up of owner-operator rigs such as this 1000-series Dodge and Fruehauf reefer trailer. This Dodge has a few extras to give it a custom look, including the deck behind the cab, which was not used for freight hauling but only for walking. At that time, the state of Missouri required one of these when traveling through the state. This picture was taken in 1982 at the Bandit truck stop in New Smithville, Pennsylvania. *Ron Adams*

Right: The A75 Autocars were always tough-looking trucks. This one has a sleeper cab and the comfort of air conditioning. The tractor is leased to L. G. DeWitt Trucking Co., Inc., of Ellerbee, North Carolina. The trailer is a Great Dane reefer. *Dick Copello* **Below:** Many refrigerated haulers came from the Lone Star State of Texas. One of those companies was Dwight Cheek Trucking, Inc., based in Amarillo. A variety of brand names made up the fleet—including this glamorous White Western Star 4800 series pulling an American reefer trailer. The cab was a 120 BBC. The tilt hood was either aluminum or optional fiberglass. *Brian Williams* **Opposite top:** This long-nose 9500-series GMC was set up as a truck-trailer-tanker unit. The popular engine for the long-nose series was the 318-8V-71. The Cummins diesel was an option. This rig belongs to Western States Petroleum, Inc., of Phoenix, Arizona. *Author collection* **Opposite bottom:** Miller Transporters, Inc., of Jackson, Mississippi, employed this R model to pull the tank trailer. Miller covered the southeastern United States and had a tanker operation with around 500 tractors and 600 trailers. Miller had 17 terminals in Alabama, Mississippi, and Tennessee. *Miller Transporters Inc., Wilbur Curtis Photography*

GMC
2RX-803
1203

MACK
Mack
8299
TRANSPORTERS
7191

U.S.
FORD
3516
4505
U.S.
VAN LINES, INC.
WORLD WIDE
YOUR BEST MOVE IS... U.S.

6847
CALIFORNIA
Y 83 259
NEVADA
51980M

Opposite top: People are always on the move, and U.S. Van Lines, Inc., of South Bend, Indiana, is there to help them do just that, whether it's nationwide or worldwide. This W-1000 Ford runs down the highways wearing America's patriotic colors of red, white, and blue. Both the dromedary and the moving van are Fruehaufs. The rig is owned by Charlie Clapp. *Neil Sherff* **Opposite bottom:** The second truck manufacturer to come out with an extended cab-over-engine was Kenworth. This cab measured 108 inches long, which was 4 inches longer than the White Freightliner Vanliner 104-inch cab that was introduced in 1968. A lot of chrome dressed up this tractor to give it an awesome look. This cab was known as the VIT The reefer trailer is a Timpte. The rig is leased in the special commodities reefer division of I-C-X (Illinois-California Express), based in Denver, Colorado. *Author collection* **Left top:** The Marmon trucks were custom built and not mass produced. This was the model HDT with the 86-inch cab owned by the Atlanta Wireworks of Atlanta, Georgia. The picture was taken in 1972 at the Union 76 truck stop in Bartonsville, Pennsylvania. *Ron Adams* **Left middle:** This CE 60 Chevrolet was powered by the big 427 V-8 engine. It was owned by Almon Bates of Oklahoma, and pulled an Americal open-top livestock trailer. The tractor, with its longer-than-usual wheelbase, has an add-on box sleeper. The wheelbases on the CE 60 series ranged from 125 to 218 inches. The standard engine was the 366 V-8. The standard transmission was a four-speed CH-465. *Harry Patterson* **Left bottom:** This 1972 Brockway K459 T had served its time, but thanks to Todd Grimm of Wynantskill, New York, it received a complete makeover. A few more chrome extras and a colorful paint job give it new life. The standard engine for the K459 T was the 8V71N 318-horsepower Detroit diesel with a five-speed transmission. This one has the 80-inch sleeper cab. The picture was taken at the 2003 American Truck Historical Society convention in Syracuse, New York. *Don Mackenzie*

Above: One of the largest trucking companies that's been around almost since the trucking industry began is Interstate Motor Freight System, Inc., of Grand Rapids, Michigan. Over the years, Interstate has had many different makes of trucks in its fleet. In the early 1970s, this Ford Louisville was one of many Fords in the fleet. It was teamed up to a Strick trailer. Interstate had many terminals that stretched from Denver, Colorado, and Texas on the western end, all the way to northern New England. They operated about 2,000 tractors and 4,000 trailers. *Interstate System* **Below:** Kenworth was always a big manufacturer of over-the-road tractors, off-highway trucks, and oilfield-rigging trucks. In 1970, the company started to get into smaller trucks and came out with the Hustler. The Hustler had many applications such as van trucks, tank bodies, block hauling, mixers, refuse, and tractors for city delivery as well as over-the-road tractors. The one pictured here is owned by Flavorland Industries. This tractor's power source was a 6172 Detroit diesel with an RT-910 transmission. The trailer is an Alloy reefer. *Kenworth Truck Co.*

Above: A new model to the Dodge's truck lineup was the CNT-950, also known as the "Big Horn." The standard engine was the Cummins diesel Super 250 with a Fuller RT-910 transmission on a 185-inch wheelbase. There were nine optional Cummins diesels and three Detroits. A sleeper box could be added with a wheelbase of 192 inches. Any length of wheelbase was available up to 292 inches and could be used in the same applications as other trucks. *Dodge Trucks* **Right:** Chevrolets of the 1970s had the same appearance as GMC trucks. This series-90 Chevrolet was leased to Steel Haulers, Inc., of Kansas City, Missouri. The driver enjoyed the comforts of air conditioning and an add-on sleeper box. There was a choice of eight diesel engines and many different transmissions. The flatbed trailer is a Dorsey. *Harry Patterson*

Right: This is a Chevrolet series-90 long-nose with a 115-inch BBC. Both Cummins and Detroit diesels were offered. The wheelbases ranged from 173 to 235 inches. This one has the optional add-on sleeper box. The photo was taken in 1980 at the Trexler fuel stop in Kuhnsville, Pennsylvania. *Ron Adams* **Below:** The White Freightliner Vanliner came on strong after it made its debut in 1968 and grew stronger over its first four years. It had 18 inches more room than the standard 86-inch cabs. John Wuethrich Creamery Co. of Greenwood, Wisconsin, hauled Grassland butter and cheese in the Timpte reefer trailer. *Harry Patterson*

Above: As we drive down the highway today, it's common to see trucks painted several shades of purple. Back in the early 1970s, it was rare to see a purple truck. This model 352 Peterbilt with a matching reefer unit to go with it on the Great Dane reefer trailer was a rare one that received attention. The truck was owned by Lowell M. Dykes and leased to Sea-Cold Services, Inc., out of Florida. The picture was taken at the Union 76 truck stop at Bartonsville, Pennsylvania. *Harry Patterson* **Below:** The long-nose 9500 GMC was popular with companies, private carriers, and owner-operators. Here we have one that's owned by Texaco. This one is wearing the standard cast spoke wheels, and the dual stacks make it a possibility that the power could be the 8V-71 318 Detroit diesel used to pull the 7,850-gallon gasoline tank trailer. The license plate shows that the rig was based in Illinois, although the location says Niles, Michigan. *Neil Sherff*

This truck is owned by the Halliburton Oil Well Cementing Co. and is a combination of the old COE-4070 and the new Paystar 5000. The new model was the COF-5070A Paystar. It had the short 50-inch cab and a GVW (gross vehicle weight) of 55,000 pounds. The load is an oil-well servicing unit. There were two engines, one in the truck and the second on the bed for high-pressure pumping action. By 1974, the Transtar II cab was used on this model and became known as the CO-5470 Paystar. *Halliburton Co.*

Another popular truck in the cab-over-engine line was the Mack F model. Many carriers favored this Mack because its short wheelbase kept it within the legal length limits. The company chose the 50-inch non-sleeper cab. Engine horsepower ranged from 180 up to 335, with transmissions from 5 to 18 speeds. *Jones Motor Co.*

Left: Pictured here is Jack Hapgood, who was president of South Loop Ford Truck Sales in Houston, Texas. The tractor shown here is a WT-9000 Ford. The new WT-9000 was a regular-looking road tractor, but it was available in an owner-operator package, which included an air conditioner, a lot of chrome items, and a choice of 12 color combinations and designs. Twelve Cummins, 1 Caterpillar, and 4 Detroit diesel engines were offered. Notice that Ford went to the small letters from the big letters across the front of the cab. *Ford Motor Co.* **Below:** Some of the trucks produced in the early 1970s are still truckin' down the highways today. This is one example of such a truck—a model 359 Peterbilt pulling a Utility reefer trailer. Many chrome extras were added to this truck and the oversize sleeper box. Extra lights complete the top and bottom of the sleeper box, as well as the step. Note the extra-long fuel tank. *Neil Sherff*

Right: This CN-900 Dodge is sitting idle waiting for a return load to go home. It's just a plain tractor with an air conditioner and an add-on sleeper box for the driver's comfort. The truck is leased to the steel division of Mason Dixon Lines, Inc., of Kingsport, Tennessee, and pulls a Trailmobile flatbed trailer. *Ron Adams* **Below:** John Lazzarin of Quesnel, Canada, chose this 5500-series GMC with the Wes Tank tanker body to deliver Gulf gas to local stations. There were three gas engines and one diesel engine to choose from as standard engines. The standard transmission was a four- or five-speed with an Allison automatic as an option. *Doug Yates Photography*

The Brute is Kenworth's contribution to on-highway construction vehicles. They claim that it is the toughest on-highway construction truck ever built. The standard engine was the NTC-290 Cummins diesel with a Fuller RT910 transmission. The front axle is rated at 18,000 pounds. The rear suspension is the Kenworth big six rated at 44,000 pounds. Optional engines included a V-6, V-8, or V-12 from 200 to 450 horsepower in Cummins, Caterpillar, or Detroit diesel. The optional transmissions were Allison automatic, Fuller, and Spicer from 4 to 16 speeds. Front-axle options ranged from 20,000 to 24,000 pounds while the rear was 44,000 to 65,000 pounds from Kenworth, Hendrickson, or Rockwell. *Kenworth Truck Co.*

The Sternlite Transportation Co. of Winsted, Minnesota, was the transportation division of the Sterners Co. A model 352 Peterbilt was chosen for mounting a dromedary body and to pull a Timpte drop-deck convertible open-top trailer. A shorter cab had to be used to accommodate the dromedary body and stay within the legal length. *Harry Patterson*

Above: The short-nose 9500 GMC has a slightly different appearance. Notice that the headlights have been moved to the lower corners of the grille instead of the center of the grille. McCormick's Express of Camden, New Jersey, teamed up this tractor with the Strick trailer. There was a choice of six Detroit and five Cummins diesel engines. The BBC was 92¾ inches. *Harry Patterson* **Below:** The Dodge "Big Horn" offered the same specifications as the other manufacturers. Under the optional equipment, any wheelbase was available, and 10 different Cummins and Detroit diesel engines were available, as were Fuller and Spicer transmissions. The powertrain in this CNT-950 is unknown, but with the 118-inch BBC the owner did not fall short on horsepower. The rig was leased to George Transfer and Rigging Co. of Parkton, Maryland. *John Hiestand*

Above: The 38 tires are checked and ready to sing on the pavement. Where else would you find a rig like this but in Michigan, the only state that allows this many axles? Fred Winkler, who owns the rig, lives in Livonia, Michigan, and the company he pulls for is Western Crude Oil in Mt. Pleasant. The sharp-looking K-100 Kenworth also has the famous Michigan-style pin-striping. The dual stacks and breather pipes indicate there could be a 12V-71 475-horsepower Detroit diesel engine powering the rig. The brand name of the crude oil trailer is unknown. Everything you see here is Michigan style. *Houghton Studio*

Below: This 9500 short-nose GMC poses nicely for picture day. Marsh Supermarkets, Inc., of Yorktown, Indiana, used short-nose tractors to pull their trailers. This tractor has the newer-style grille with the headlights at the bottom rather than in the center. Dual stacks beef up the appearance. *Author collection*

Above: Three tanks make up this combination. The aluminum tanks that are pulled by the White Freightliner truck are hauling crude oil, burner fuel, and asphalts. W. S. Hatch Co. of Woods Cross, Utah, is the tank carrier that owns the rig. Hatch covers most of the western half of the United States. Hatch had tank trailers for almost every type of bulk hauling. *W. S. Hatch Co.* **Right:** FWD Truck Corp. made trucks mainly for construction work, but it also made some over-the-road tractors. Here we see an FWD double take. The tractor is an FWD model CO5-2178 cab-over-engine with an FWD fire apparatus truck on an outside frame flatbed trailer. By looking at the number on the tractor, FWD must have had several of these tractors in their fleet. *Ron Adams*

Above: Another truck that has been brought back to life is this 1972 9500 long-nose GMC. George Brown of Hillsboro, Tennessee, did the restoration work and did it well. Most of the long-nose GMCs had the 8V-71N V-8 Detroit diesel engine. The only wheels available were cast spokes. *Dick Copello* **Left:** The series 90 was available in both short- and long-nose. What we see here is the short-nose version pulling Fruehauf tri-spread axles. The rig is owned by the Chevrolet division. The short-nose had a 93-inch BBC and a GCW of 76,800 pounds. The wheelbases ranged from 146 to 210 inches. V-6 or V-8 gasoline and Cummins and Detroit diesel engines were available. With the dual stacks and the tractor being a General Motors product, the engine is most likely a Detroit diesel 8V-71 with 318 horsepower. *Joe Wanchura*

MO. 13610
OKLA. 26527
TENN. HF 6693
ALL POINTS, Inc.
TRUCK RENTALS
MIAMI, FLA.
A 224

EBY BROS
17
17

Opposite top: In the early 1970s, Marmons were still relatively new trucks on the market. Because Marmons were custom built, sales were not quite as high as those of their competitors. This model HDT-BC-86 had the 86-inch cab. The standard engine was the Cummins NTC-350 with high-mount fan. The Florida palm trees make a nice background for this Marmon owned by All Points Truck Rentals in Miami, Florida. *Author collection* **Opposite bottom:** The patriotic colors of red, white, and blue finish off this K-100 Kenworth very nicely. The standard engine and transmission were the NTC-290 Cummins diesel with the Fuller RT910 10-speed transmission. Optional engines were Cummins with 16-horsepower ratings from 190 to 400, six Detroits from 195 to 475 horsepower, and three Caterpillars from 270 to 425 horsepower. Five cab sizes were available from 52 to 86 inches. Eby Brothers of Boise, Idaho, chose a long wheelbase for mounting a dromedary body on the chassis to haul extra freight with the flatbed trailer. The load is 125 10-foot pieces of galvanized pipes. In a lot of cases, the money made from the freight on the dromedary body pays the trip expenses, while the rest of the load is money in the bank. *Harry Patterson* **Above:** Since the exact model number of this truck is unknown, the powertrain in this 400-series tractor can't be identified. Engine choices included Cummins, Caterpillar, and Detroit diesels with horsepower ranging from 230 to 434. Transmissions were Fuller and Spicer. This tractor was fitted with the 86-inch sleeper cab. Zurn Industries, Inc., of Erie, Pennsylvania, owned the Brockway tractor and Great Dane trailer. *Author collection* **Left:** This brand-new GMC 6500 is on its way for delivery to Nabisco. Nine different gasoline engines with horsepower from 170 to 275 and two diesels at 165 and 195 were available. The cab is a 72-inch tilt. The transmissions were a four- and five-speed and an Allison automatic. The photo was taken in 1973 at the Union 76 truck stop at Bartonsville, Pennsylvania. *Ron Adams*

Top: Here we have a Mack FS-700 western 50-inch cab pulling a set of Monon trailers. The model was either an FS (steel) or an FI (aluminum). Advance-United Expressways, Inc., of Minneapolis, Minnesota, owned the rig. *Author collection* **Above:** Here we have another rig for Advance-United Expressways Inc. This is a U-model Mack pulling a set of Monon trailers. The U model was the offset cab. The U 600 offered the ENDT 673 six-cylinder, a 675-676 Maxitorque, or an ENDT 865-866 Maxitorque with horsepower from 180 to 375. The companion to the Maxitorque engine was the Maxitorque transmission. *Author collection*

Above: Hill & Hill Truck Lines, Inc., of Houston, Texas, was a heavy-hauler carrier of oversized and oil-field equipment. In this business, various types of trailers with high weight capacities used to meet customers' needs. Pulling this low-bed trailer is a White Western Star. It has a 120-inch BBC with a 90-degree tilt hood and is powered by a Caterpillar diesel engine. A sleeper box was needed for those many nights away from home. The new 1,445-square-inch radiator was now standard with the Cummins NTA-370 diesel engine. *Harry Patterson* **Left:** The top photo shows one of the first tractor-trailer mixer applications. It was hooked up to a 1915 International model H featuring a four-cylinder engine, three-speed transmission, and a 1,500-pound capacity. The bottom photo is a Paystar F-5070 that shows a lot of big improvements and developments over a half century. It's powered by a Cummins NTA-370 horsepower diesel engine with a 16-speed transmission and has a GVW of 85,000 pounds. The mixer body is a Challenge. *International*

Above: Many loads of cheese and dairy products were shipped to the Southwest and Pacific coast. One of the carriers for these was Pirkle Refrigerated Freight Lines Inc. of Madison, Wisconsin. The company was started by Otto Pirkle around 1950. Its early ventures were to the West Coast and eventually came to the East Coast. The fleet was made up of owner-operator equipment such as this model 352 Peterbilt and Timpte reefer trailer. Timpte trailers came on real strong through the 1970s. *Harry Patterson* **Below left:** Another refrigerated carrier that hauled swinging beef and frozen meat products was Texas Continental Express Inc. of Eulis, Texas. This fleet was also made up of owner-operator equipment. One of those rigs was this GMC Astro 95 and Timpte reefer trailer. The picture was taken at the Union 76 truck stop in Bartonsville, Pennsylvania, in 1973. *Bob Parrish* **Below right:** From Pocono country comes Paul Ace Trucking, Inc., of Stroudsburg, Pennsylvania. Paul Ace owned a fleet of trucks that was made up of several brand names of trucks, including this 360 Brockway. Ace was leased to Creif Bros. Corp. also of Stroudsburg, Pennsylvania. Paul Ace went the extra mile to make his equipment look impressive, which made not only a good impression on his company, but also on him. This Brockway is powered by a Cummins diesel. *Harry Patterson*

Above left: This model HDT-BC-86 was owned by McKee Baking Co. of Collegedale, Tennessee. This model was powered by a Cummins NTC-350 diesel engine and a 13-speed Fuller transmission. The trailer is a Brown reefer. *Harry Patterson* **Above right:** Moving companies were big users of dromedary rigs. Aero Mayflower Transit Co., Inc., of Indianapolis, Indiana, was no exception. The tractor was a Chevrolet Titan 90 and pulled a Fruehauf moving van trailer. The picture was taken at Truck Stops of America in Frystown, Pennsylvania, in 1973. *Ron Adams* **Below:** The Kenworth C 500 was designed for several applications. The J & H Logging Co. of Petersburg, Alaska, put this C-500 to work hauling logs. The standard engine was the NTC-290 Cummins diesel and the RT-910 Fuller transmission. The wheelbase depended upon the application. *Kenworth Truck Co.*

GMC
NUTRA-FLO

WHITE FREIGHTLINER

Opposite top: Not much is known about this photo except that the tractor is a GMC Astro 95 and the unknown brand trailer is for liquid petroleum and goes by the trade name Nutra-Flo. *Author collection* **Opposite bottom:** By this time, 42-foot trailers were being used and drivers were starting to paint the trailers to match the tractors. C. Grant of Los Angeles, California, was the owner of this White Freightliner and Brown reefer trailer. Air scoops and covers for reefer units were being implemented for fuel consumption because of the fuel crunch of 1973. White Freightliner offered cabs from 48 inches up to the super-size 104-inch Vanliner cab. *Harry Patterson* **Top left:** Autocar trucks were always tough-looking, but add chrome and other extras and they looked beautiful. A good example is this DC 10364. It is owned by Sasso Contracting Co., Inc., in Newark, New Jersey. It pulls a lowboy trailer and is dressed in its Sunday best. The power source is unknown, but I'm sure that the right engine was selected for the job. Although it is not visible in this photo, the truck nicknamed *Screaming Eagle* is set off in light green pinstriping. *Harry Patterson* **Top right:** In some cases, there are things that make some trucks stick in your memory for certain reasons. In this case, the memory is the sound that came from that stack. I live along Interstate 78, and one morning as I was going to my car to leave for work, I heard the sound of a truck coming up the highway. I heard this sound more than a mile away, but waited to get in my car until I could see what kind of a truck it was. It was a W-1000 Ford. The sharp cracking sound that came from the stack of this truck was awesome and unreal. The truck was owned by Peacock Trucking of Lovettsville, Virginia, and pulled a Hobbs Vanguard livestock trailer. The picture was taken in July 1973 at Truck Stops of America in s Frystown, Pennsylvania. *Ron Adams* **Left:** Sunday afternoons and nights through Pennsylvania is always a heavy traffic time for trucks to make their Sunday night and Monday morning deliveries. One of the trucks in that heavy Sunday truck traffic was this White Western Star and Utility reefer trailer that was leased to National Carriers Inc. based in Liberal, Kansas, who was a nationwide refrigerated carrier. The tractor does not have much chrome, but still, it's a great-looking truck. The picture was taken on a late Sunday afternoon in October 1973 at the Union 76 truck stop at Bartonsville, Pennsylvania. *Ron Adams*

Above: The Cargostar received a face-lift with a different grille and bumper in 1973. It had an all-steel tilt cab. The standard engine was the International VS-401 V-8 gasoline. Five other gasoline engines were offered: RD-406 160 six-cylinder, RD-450 168 six-cylinder, RD-501 181 six-cylinder, VS-478 209 V-8, and VS-549 227 V-8. The standard transmission was a T-54 five-speed. Eleven other transmissions were offered as optional, including four- or five-speed Allison automatics. Wheelbases ranged from 125 to 192 inches. The cast spoke wheels were standard. *International* **Below left:** North American Van Lines of Fort Wayne, Indiana, had all makes of tractors in its fleet. This Chevrolet had a sleeper cab and probably a 6V-71 diesel engine. What's a little unusual about this rig is that almost all the moving van trailers had tandem axles; this Fruehauf is still a single axle. The picture was taken in July 1973 at Truck Stops of America at Frystown, Pennsylvania. *Ron Adams* **Below right:** This rig is typical of one of the different types of trucks that operated in the west. Yandell of Oakland, California, operated this FL-700 Mack with a flat dromedary body pulling a flatbed trailer with an undercover load. This one has the 80-inch sleeper cab. *Author collection*

Above: Bray Lines, Inc., of Cushing, Oklahoma, was a refrigerated carrier that hauled foodstuffs, beverages, frozen products, feed, petroleum products, glass, carpets, sugar, salt, and photographic equipment in the continental 48 states. A fleet of 250 tractors with 350 trailers was used to get the jobs done. One of those trucks was this White Freightliner with the 104-inch super-size cab, known as the Vanliner. A cab like this came in handy when making long trips away from home. It was nice for a husband-and-wife team because two could sleep in the double bunk. Mr. and Mrs. Gregory enjoyed their trips in this colorful cab. *Harry Patterson* **Left:** This Paystar 5000, or what was the 5070 6x6, proved what it could do on the day of this photo shoot. I'm sure it had no problem crawling out of this hole. The standard engine was a Detroit 6-71N diesel with a Fuller T-459 13-speed transmission. The front axle was rated at 16,000 pounds, while the rear was rated at 38,000 pounds. The cab was a 114-inch BBC. Disc wheels were standard while cast spokes were optional. The wheelbases ranged from 182 to 212 inches. *International*

Above: Kenworth was the second manufacturer to come out with a super-size cab. This cab, known as the Vit, measured up to 108 inches—4 inches more than the 104-inch Vanliner. It was the ideal cab for the husband-and-wife team of Charles and Ruth Dohn. They leased their rig to Mason Dixon Lines, Inc., of Kingsport, Tennessee, traveling cross-country hauling acids, dry and liquid chemicals, syrups, feed, fertilizer, petroleum products, edibles, and other products. Mason Dixon Lines was formed in 1936. *Harry Patterson* **Below:** Peterbilt was the next company to introduce a super-size cab at 110-inches. It offered basically the same interior package as Kenworth and White Freightliner. This was available on the model 352. Hagen, Inc., had the honors of hosting this nicely painted tractor to pull cross-country for them out of Sioux City, Iowa. Frozen and refrigerated cargo were its main hauls. *Harry Patterson*

Top: White Western Stars did all kinds of hauling. This truck pulled a Polar tank trailer. Most of the White Western Stars wore the dual headlights, but this one had singles. *Author collection* **Above left:** Many drivers liked the Diamond REO C-11664, including the owner of this truck who leased his rig to Riss & Company of Kansas City, Missouri. On this model, the stacks could be mounted in two positions: in front of the sleeper box at the cab or behind the sleeper box as shown here. The air conditioner kept the driver cool as he rode down the highway. The trailer is a Strick. The picture was taken in the summer of 1973 at Truck Stops of America at Frystown, Pennsylvania. *Ron Adams* **Above right:** This big Brute Autocar DC9364 was owned by an excavating company in Linden, New Jersey. The chrome extras, such as the bumper guides, fender mirrors, air horns, and grille guard give this Autocar a good-looking appearance. The power source is unknown, but the dual stacks could represent a Detroit diesel. *Harry Patterson*

Above: The Brockway 457 TL was on a 142-inch wheelbase. The horsepower ranged from 230 to 370 with a 5- to 16-speed transmission. The rear-axle rating was 34,000 to 44,000 pounds with a gross weight rating of 73,000 pounds. The engine choices were Cummins, Caterpillar, and Detroit diesels. The cab was the 86-inch BBC. This tractor has the optional disc wheels. *Brockway Motor Trucks* **Below:** This White Freightliner–named Black Beauty–is leased to Hill & Hill Truck Lines, Inc., of Houston, Texas. Heavy hauling and oil field hauling are the rig's main duties. Chrome items and pinstriping dress up the cab. A dromedary body was added to haul extra freight. The Fruehauf flatbed trailer is loaded with lumber. Although Black Beauty is not the source of horsepower, there are more than enough horses under the cab to get the job done. *Harry Patterson*

Above left: From northern New England comes Coles Express, Inc., of Bangor, Maine. A fleet of 130 tractors and 300 semi-trailers served the New England area and New Brunswick and Nova Scotia, Canada. Coles was a long-time user of Diamond T and its Diamond REO trucks. This tractor is a Diamond REO Royale with a 54-inch BBC day cab. It has a 90-degree tilt cab. Available engines were Caterpillar, Cummins, and Detroit diesels up to 475 horsepower, although the power in this tractor was not that high. The trailer is a Strick. *Bob Parrish* **Above right:** Autocar trucks were used not only in construction and as over-the-road tractors but also in recovery. This DC75 was owned by Ted Covington of Ottsville, Pennsylvania. It rode on the standard cast spoke wheels. *Harry Patterson* **Below:** There were many refrigerated haulers that hailed from the Lone Star State of Texas, including Jay Lines, Inc., of Amarillo. It primarily hauled meat and frozen food products. Its fleet was made up of owner-operators who had eye-catching equipment. One of those tractors is this fancied-up model 352 Peterbilt that pulled a Utility reefer trailer. In this era four- and five-color paint schemes were common on many tractors compared to today's many single colors. The owner of the tractor is unknown, but whoever he may be, he certainly made a good choice. *Harry Patterson*

This LT-9000 Ford was used to haul logs in the lumber industry. Little is known about the truck itself except that it is owned by G. Dickie & Son of Neopit, Wisconsin. The cab is a 105.3-inch BBC. GVWs ranged from 22,100 to 61,000 pounds with GCWs to 80,000 pounds. Power sources were up to 534-cubic-inch gas engines and 927-cubic-inch diesels. According to the 1973 detailed component descriptions book, there were 14 engine choices, 2 standard and 12 optional, and 20 main transmissions available, including Allison automatics, Clark, Fuller, and Spicer. *Author collection*

The White Western Stars proved themselves to be trucks worthy of being in any field of hauling. Gulf Oil used this truck-trailer based in Portland, Oregon, to deliver gas to local gas stations. Tanker truck–trailer combinations were one of several different ones used in the western states. Both trucking companies and private carriers used them, as well as owner-operators. *Photo-Art Commercial Studios Portland, Oregon*

This C65 series Chevrolet was employed by Allied Van Lines, Inc., of Broadview, Illinois. The standard engine was a 366 V-8 with four- or five-speed transmissions from Chevrolet, Allison, New Process, Clark, and Spicer. The wheelbases ranged from 125 to 218 inches with GVWs from 21,000 to 40,500 pounds and a GCW of 60,000 pounds. The 97 1/2-inch BBC cab is dwarfed by the extra-high sleeper box. The trailer is a Fruehauf moving van. *Harry Patterson*

The International 4370 Transtar was competitive with other over-the-road conventional tractors. The power source in this tractor is unknown, but the little round sticker on the side of the hood signifies that it is equipped with a Jacobsen engine brake, known as a Jake Brake. The rig is leased to Coast Provision Co. of Santa Fe Springs, California. The cab is a 117-inch BBC with a 60-inch sleeper box. The wheelbases ranged from 160 to 240 inches. The standard engine in the 4370 was the NTC-290 Cummins with an RT-910 10-speed transmission. There were four optional Cummins diesels: super 250, power torque 270, NTC-350, and NTZ-370. Transmissions from 4 to 16 speeds were optional. *Harry Patterson*

The Hustler was a low-profile cab-over-engine model from Kenworth. City delivery work was the idea behind the Hustler. The horsepower choices ranged from 210 to 262, and 10 different transmissions were available. Single-drive axles went from 18,000 to 22,000 pounds, while the dual-drive axles went from 34,000 to 44,000 pounds. Several applications were available for the Hustler. This tanker, made by Beall Trans-Liner, was owned by Olympic Oil Co. *Beall Trans-Liner Inc.*

CHAPTER 2

1974–1976

In 1975, Brockway discontinued the 300 series and replaced it with the new 700 series. This is when Brockway started using the Mack R-series cab. The 759 was a 90-inch BBC with the high-mount hood and cab. The 760 was a 117- or a 121-inch BBC axle-forward model. The 761 was a 121-inch BBC axle-back model.

Diamond REO went out of business in 1976, and Dodge ceased producing trucks in 1975. Two good truck manufacturers are now in the pages of history.

In 1951, Freightliner and the White Motor Co. signed a sales agreement to market Freightliner trucks at White dealerships, provided the name *White* would go on the Freightliner nameplate to make it White-Freightliner. The agreement was signed for 25 years. In 1976, the sales agreement expired and the name changed from White-Freightliner back to Freightliner. Freightliner went out on its own and was a great success. The conventional model FLC-12064 had a 120-inch BBC cab. The cab-over-engine models offered 63-, 75-, 86-, 96-, and 104-inch BBC cabs. The Powerliner, with its massive grille, could handle engines with up to 600 horsepower.

The International Transtar 4100 Conco was introduced in 1974. It was a tractor designed for both

A new model from Mack was the Cruiseliner. It had the biggest cab of the standard cabs. The cab we see here is the 54-inch BBC. It's also available with the 76- and a 90-inch sleeper cab. It sports a larger radiator that can accommodate engines up to 500 horsepower. It had a selection of 28 engines with horsepower ranging from 237 to 500. A wide range of transmissions was also available. A long wheelbase accommodates a dromedary body due to the light weight of the cargo being hauled for more payload. Holsum Bakers owns the rig. *Author collection*

Hauling heavy machinery and oilfield equipment requires a tough and rugged truck. J. H. Rose Truck Line, Inc., of Houston, Texas, felt that this Autocar would fit the bill. The oversize bumper and grille, along with the construction fenders, made this truck ready for the job. It's powered by a Caterpillar diesel and has an air conditioner for the driver's comfort. The photo was taken at the Cheyenne, Wyoming, port of entry at Interstate 80 and 25. *Harry Patterson*

pickup and line-haul work. It had a 66-inch BBC and was offered in both tandem- and single-axle versions. The S series replaced both the Loadstar and Fleetstar model S. The S series offered a 91-, 111-, and a 112-inch BBC for the 2200, 2500, and 2600 models, respectively. The S series that replaced the Loadstar included models 1600, 1700, 1800, and 1900 with a 97.5-inch BBC, and the 2100 had a 91-inch BBC. International trucks were powered by International's own engines or Cummins, Caterpillar, or Detroit engines.

Kenworth introduced a new cab length of 108 inches in 1974. This was Kenworth's new Vit cab with a much larger bunk.

Peterbilt came out with 11 new models in the 1970s, and only one truck was made of the 248 and 253 in 1976. The highest production model was the 352 with almost 29,000 units produced.

The fuel crisis of the mid-1970s led to the design of air foils and air scoops for better fuel economy, which eventually created the evolution of aerodynamic trucks.

The Chevrolet Titan 90 was the same as the GMC Astro 95, except for a different name. Available engines included six Detroit diesels, four Cummins diesels, and the 3406 Caterpillar. Various transmissions were also available. This sharp-looking Titan 90 is set up as a truck-trailer combination. It has the 86-inch sleeper cab with extra chrome and the comfort of an air conditioner. Both the body and trailer are Comet. *Author collection*

Triples were an experiment in the 1960s, and they became legal in the 1970s. Garrett Freightlines Inc., of Pocatello, Idaho, was one of the many companies in the West that used them. The tractor is an International Transtar 4270. All three trailers in the triples are Comets. The license plate reveals that this photo was taken about 1974–1975. Garrett Freightlines, Inc., covered 11 western states and into Minnesota in the northern states. *Garrett Freightlines Inc.* **Opposite:** Hayes Truck Manufacturing Ltd. was owned by Mack Trucks in the 1970s. This handsome 1974 Hayes Clipper was owned by Trevor Cameron of Alberta, Canada. Although no specs are known about the truck, it's likely that Canadian trucks had the same engines, transmissions, and other components as U.S. trucks. There was no shortage of chrome on this Hayes. *Don Mackenzie*

BrockwaY
Diesel
GENERAL

RUNNIN' BARE

Opposite top: This Brockway model 457T could be mistaken for a Mack from a distance. At close range, it is definitely recognizable as a Brockway. The reason they look so much alike is because the 400 series is actually a Mack cab. The 457T horsepower ranged from 230 to 290, with transmissions ranging from 5 to 16 speeds with a 73,000-pound gross weight (GW). The wheelbase was the standard 140 inches. Brockway started to design its own custom color schemes and paint designs. *Brockway Motor Trucks* **Opposite bottom:** This Diamond REO C11664 was one of several trucks in the fleet of Arthur Shelley, Inc., of Dallas, Pennsylvania. Diamond REO offered the Cummins, Caterpillar, and Detroit diesels as the power source with a number of transmissions to choose from. *Runnin' Bare* did not have much chrome, but it was still a nice-looking truck. *Harry Patterson* **Above:** The White Western Star was only seven years old in 1974. It had come a long way in such a short time. This tractor is leased to Delta Truck Lines Ltd. of Toronto, Ontario, Canada. Most of the time you can identify Canadian tractors by the spread of the tandem. This one has the 120-inch BBC. Notice the extra-long fuel tank. The paint scheme is one of White Western Star's own designs. *Harry Patterson* **Right:** Because Marmons were custom built, they didn't have the large production runs their competitors did, but Marmons were as good as competitors' trucks and got the job done just as well. The standard engine was the big Cummins NTC 350 with the Fuller RTO 13-speed transmission. The 86-inch cab was of aluminum construction. This model HDT-BC-86 was working for KSI Farm Lines, Inc., of Wisconsin and Los Angeles, California. *Harry Patterson*

Right: The GMC Astro 95s worked in all 48 states. This outfit was leased to Indian Hills Trucking Co. of Great Falls, South Carolina. Its job was hauling produce and other refrigerated commodities in the Timpte reefer trailer. This Astro 95 was powered by a Cummins diesel. *Harry Patterson*
Below: McLean Trucking Co., Inc., of Winston-Salem, North Carolina, was one of the largest trucking companies in the country. The fleet numbered more than 5,000 tractors and 9,000 trailers, including these White Road Commander tractors that could pull a 45-foot trailer and be within the 55-foot legal limit. McLean had 216 terminals in 43 states and British Columbia. *McLean Trucking Co.*

Above: By 1974 and 1975 the International Transtar II took on a different appearance. Two of the changes were the chrome bowl around the headlights and the chrome side plates on each side of the grille. The standard cab was a 55-inch BBC. Optional cabs were the 74- and 84-inch BBC sleeper cabs. The transmission was a Fuller RT-910 10-speed. The specs about this particular truck are unknown, but it has the 84-inch sleeper cab. This fancy tractor is owned by the Georgia-Pacific Corp. of Tigard, Oregon, and pulls a Trailmobile trailer. The four men in the picture are unidentified. *Georgia-Pacific Corp.* **Left:** Here we have a Brockway model 457 cab-over-engine tractor. This photo was taken on the used truck lot of Horwith Freightliner, Inc., in Kuhnsville, Pennsylvania, in 1980. No specs are known, but it is a great-looking 457. *Ron Adams*

Gulf
M·C
TRANSPORTATION
Worland & Laramie
ICC-MC59436
COLO-RO-31237
MONT G.V.W. 44000
EMPTY WT 15800

LATEX
NONFLAMMABLE
ICC MC 30887
SHIPLEY
TRANSFER INC.
REISTERSTOWN-MD.
502
73.280

Opposite top: The model 352 Peterbilt was available in five cab sizes: 54, 63, 76, 86, and 110 inches. Engines such as Cummins, Caterpillar, and Detroit were available with various horsepower ranges and transmissions in various speeds. The cab selection we see here is the 110-inch version. The owner is M-C Transportation of Worland and Laramie, Wyoming. Its operation was livestock, sand and gravel, and heavy hauling. In this case, it happens to be a livestock rig pulling a Wilson potbelly trailer. *Harry Patterson* **Opposite bottom:** A nice, clean tractor always makes a good impression on the driver and the company it represents. Such is the case with this White Freightliner leased to Shipley Transfer, Inc., of Reisterstown, Maryland. With a rig like this, the driver is proud to drive through the 48 states that Shipley serves in its tanker operation. The gross weight is 73,280 pounds. *Harry Patterson* **Left top:** One of the big stars in the International line was the Fleetstar. This F-2070A was set up as a truck-trailer combination, with a 6-71N Detroit diesel engine and a Fuller RT-910 10-speed transmission as standard. It had a 92-inch BBC with a wheelbase range from 142 to 220 inches. Both the body and trailer are used for hauling milk and are made by Polar Tank Co. *Author collection* **Left middle:** Here we have a White Western Star taking a break at the Jersey Truck Center. It was leased from the steel division of Mason Dixon Lines, Inc., of Kingsport, Tennessee. Mason Dixon Lines was one of many common carriers that had a steel division and a tanker division. Not many conventional tractors at this time had the stacks mounted behind the sleeper box. The location of the battery box and fuel tank is the reason for the location of the stacks. *Harry Patterson* **Left bottom:** The Road Boss was a popular tractor for White. There were three BBC choices: 92, 108, and 122 inches. What we see here is probably the 108-inch version. Although the name on the door shows Chemical Leaman Tank Lines, Inc., of Downingtown, Pennsylvania, the truck is owned by Gotwal Bros., Inc., of Logansport, Indiana. It was leased to Chemical Leaman and based at the Evansville cement plant in Evansville, Pennsylvania, and hauled dry cement. The picture was taken in October 1974 in Kutztown, Pennsylvania. The truck was driven by Terry Bailey of Kempton, Pennsylvania. *Ron Adams*

Above: This W-900 Kenworth is from around 1973-1974. By this time, the W-900 had been in production for about 15 years, and because of continuing demand the W-900 is still in production today. Apache Tank Lines owns this Kenworth truck-trailer-tanker rig that is leased to the Cowboy Oil Co. of Pocatello, Idaho. Although no specs are known about this rig, it is equipped with a Jacobsen engine brake (Jake Brake). These truck-trailer combinations were very popular in the west for many years. *Author collection*

Right: When you look at the International Transtar 4100 Conco it gives the impression that you are looking at a 1950s RDCO-405 cherry picker. It had a 66-inch BBC, which gave it the authority to pull a 45-foot trailer within the legal length limit. It could also pull 27-foot doubles in states that allowed 65-foot rigs. The cab sat lower, farther back, and offset and could still accommodate a big 350-horsepower diesel engine. The one shown here is a restoration piece from 1975 and is owned by K & M Enterprises of Perrysburg, Ohio. *Don Mackenzie*

Above: White Freightliner had the first extra-long cab at 104 inches—known as the Vanliner in 1968. Just five years later, the big-grille Powerliner was introduced. It had a big 2,000-square-inch radiator, which meant that it could handle a 600-horsepower engine. This Powerliner was leased to Badger Freight Service of Cudahy, Wisconsin, which was a division of Indiana Refrigerator Lines, Inc., of Muncie, Indiana. *Harry Patterson* **Below:** A lot of Canadian rigs had their own setup and axle spreads like those in Michigan. Donline Haulage Ltd. owns this unusual rig. The tractor is an RL-700 Mack. Its power comes from the ENDT-865 V-8 Maxidyne engine that delivers 325 horsepower and was more than likely teamed up with the Maxitorque transmission. Cummins, Caterpillar, and Detroit diesel engines were also available with up to 475 horsepower, along with Spicer and Fuller transmissions with up to 18 speeds. The long-frame dump trailer is a Trailmobile. *Micklethwaite*

Left: When it comes to customizing, Marmons can look just as good as any other truck. This 1974 model GHD is owned by Virgil Hawkins of Dunnellon, Florida. The 350 Cummins and 10-speed Fuller transmission are both between a 255-inch wheelbase. Single axles are rather unusual on wheelbases of this length. All the extra chrome makes this truck a real eye-catcher. *Harry Patterson* **Below:** Time DC is the combination of three trucking outfits that merged in 1969. The companies were Time Freight, Inc., of Lubbock, Texas; Los Angeles-Seattle Motor Express, Inc., of Seattle, Washington; and Denver-Chicago Trucking Co., Inc., of Denver, Colorado. Time Freight and Denver-Chicago were had already been coast-to-coast carriers since 1962 and 1946, respectively. Los Angeles-Seattle (Lasme) had been a Pacific coast carrier since the early 1930s. All three carriers ran double trailers in their operations and continued to do so after the merger. White Freightliner tractors and Strick trailers helped make up the fleet of more than 800 tractors and 3,500 trailers serving 80 terminals. The company went out of business around 1981. *Time-DC*

Left: The International 4100 Conco was used widely across the country. One of the companies that used the Conco was Garrett Freightlines, Inc., of Pocatello, Idaho. This truck was used to pull triple trailers in states where they were allowed. The head-on appearance gives it almost the same look as the RDCO-405 of the 1950s. Clarence and Oscar Garrett started Garrett Transfer & Storage Co. in 1913 with a REO truck. They later changed the name to Garrett Freightlines, Inc. *Author collection*
Below: One of the companies that has had great success over the years is C. R. England & Sons, Inc., of Salt Lake City, Utah. Produce and refrigerated commodities were its main hauls. In 1963, I saw one of its narrow-nose Kenworths come into the produce center in Philadelphia with a load of California iceberg lettuce from the Salinas Valley. The trip was made in 52 hours with three drivers. Their comment was, "You're not going to get iceberg lettuce any fresher than this from California." England is well-respected for its clean, sharp-looking equipment. This K-121 Kenworth is a good example, along with the set of Trailmobile trailers. England is still operating today. *Author collection*

Great-looking trucks always catch a shutterbug's eye, and Peterbilt 359s were no exception. This one has plenty of chrome and polished aluminum wheels all the way around, even on the Utility reefer trailer. I can't identify the company, but the truck is based in La Mirada, California. The 359 has a 113.25-inch BBC. The standard engine was the Cummins NTC-350 with a Fuller RTO 13-speed transmission. Other optional engines were Caterpillar and Detroit diesels with a wide range of power up to 600 horses. Allison and Spicer were optional transmissions. The 359 also had a 119.25-inch BBC, depending on the application. *Harry Patterson*

The International Fleetstar 2070-A was a good truck for several applications. The standard engine was the 6-71N Detroit diesel with the RT-910 10-speed transmission. It had a 92-inch BBC. Wheelbases ranged from 142 to 220 inches. There were seven optional engines with horsepower up to 335. The Fleetstar series was discontinued in 1976 and replaced with the new S series. *International*

Above: This Kenworth W-900 Vit is quite photogenic. It has a 60-inch sleeper box that was replaced with the Aerodyne sleeper a few years later. There was plenty of room in the sleeper for extras like a coffeepot, television, cooler, and tape deck. The standard engine was the NTC-290 Cummins diesel with a Fuller RT-910 10-speed transmission. Optional engines were Cummins, Caterpillar, Detroits, V-6, V-8, and V-12 up to 450 horsepower. Spicer, Fuller, and Allison transmissions with 4 to 16 speeds were optional. Both the tractor and Fruehauf trailer must have been brand new when this photo was shot because no identification is available. *Neil Sherff* **Opposite bottom:** The first prototype of the present-day Freightliner was produced in 1973 and went into regular production in 1974. The engine was a Cummins NTC-290. The new conventionals wore the White Freightliner name, but this one is only wearing the Freightliner name. Notice that this new conventional has three steps to enter the cab—two on the fuel tank and the third at the bottom of the cab. It had a 120-inch BBC with the same engines and transmissions as the cab-over-engine models. A chrome and custom package was also available for those who wanted to fancy up their rig. *Freightliner Corp.*

Above: White Western Star trucks applied themselves in all fields of trucking as their competitors did. The 120-inch BBC was one of the longest in the industry. It sits on a long-wheelbase chassis to accommodate the long body and is set up to pull the little tote-along trailer. The front axles were rated at 12,000 pounds, while the rear tandem axles were rated at capacities from 32,000–50,000 pounds. Builders Supply of Fargo, North Dakota, is the owner. *Author collection* **Right:** Double trailer operations came in two types: 26- and 27-foot and 40-foot-plus doubles. What you see here are the doubles of the 40-foot types. The tractors, from left to right, are a White Road Commander, a Louisville Ford, and another Road Commander. The equipment was owned by Red Star Express Lines, Inc., of Auburn, New York. These types of combinations were run on the Ohio, Indiana, and Massachusetts turnpikes and the New York State Thruway. Red Star only ran them in New York and Massachusetts. The Red Star fleet totaled over 800 tractors and 1,500 trailers. *Author collection* **Opposite top:** In late 1973, Brockway introduced the new 700 series. One of the new models was the 761, which is the axle-back version. The 121-inch BBC was on a 183-inch wheelbase. Available engines were Cummins, Caterpillar, and Detroit diesels matched to a Fuller RTOF 13-speed transmission. The cab is a model 78, which is the same cab that Mack used on its R-series trucks. Dana, Allison, and Cummins Sundstrand automatic transmissions were optional. Some optional engines, such as the 8V-92 Detroit diesel, offered as much as 430 horsepower. The Huskie series of trucks were real workhorses. *Brockway* **Opposite bottom:** The constructor model of the Fleetstar F-2070A is a great-looking unit. It's equipped with a TI-Brook 15-cubic-yard dump body. The Detroit diesel 6-71N with 238 horsepower was the standard engine. The highest optional horsepower was 335 from the NTC-335 Cummins diesel. The gross vehicle weight was 60,000 pounds with a gross capacity weight of 79,000 pounds. The standard transmission was the T-135 5-speed with options up to 13 speeds. The color is robin egg blue. *International*

Brockway
Brockway
GENERAL
GTX

INTERNATIONAL

Left: The Marmon CHDT-BC was as versatile as any other truck in the industry. The standard engine was the Cummins NTC-350 high-mount fan diesel with a Fuller RTO-9513 13-speed Roadranger. The radiator had a 1,200-square-inch frontal area. The driver had the comfort of an air conditioner and a sleeper box. *Harry Patterson* **Below:** The Dodge CNT-900 was a dual-drive tandem-axle tractor. The engine was the Cummins NH-230 or optional NHC-250 with a Spicer short fourth five-speed transmission. The GVW was from 41,000 to 53,000 pounds with a GCW to 76,000 pounds. The wheelbases ranged from 146 to 200 inches with a 90-inch BBC. *Dick Copello* **Opposite top:** Owner-operators accepted the new White Freightliner conventional with no problem. As we see here, you could dress them up to look nice. This was strictly a western rig because of the long wheelbase and the oversize add-on sleeper box. They were not allowed in the East. The company is unknown. The Trailmobile trailer hooked to the tractor to make a nice combination that rode on polished aluminum wheels all the way around. *Harry Patterson* **Opposite bottom:** Kenworths have always been a good target for lots of shining chrome and fancy paint schemes. This one is a good example of that. Road Runner Trucking, Inc., of Omaha, Nebraska, whose fleet was made up of owner-operators who had many nice trucks like this one. The W-900s were very popular with owner-operators. The Timpte reefer trailer makes a nice match for the tractor. The rig was operated by Beall Trucking of Fremont, Nebraska, and leased to Road Runner Trucking. *Neil Sherff*

THERMO KING
103
Runner
LARSEN
FREMONT NEBR

The Diamond REO Raider was the next truck to come from Diamond REO. It was available in two versions: the setback front axle and the axle forward. Both had a 119-inch BBC. It had a 1,600-square-inch radiator to handle all sizes of engines, including the KT-450 and 8V-92T. Cummins, Detroit, and Caterpillar were the selections. Fuller and Spicer manuals and Allison and Cummins Sundstrand automatics were the transmission choices. Aluminum was standard on the axle-forward models and optional on the axle-back models. Ed Kemp of Pierpont, Ohio, chose the axle-forward model with the 36-inch sleeper box. He leased his rig to All States Trucking of Oakland, California, which was the reefer division of P-I-E. *Harry Patterson*

Can you tell which state this rig is from? If you guessed Michigan, you're right. This type of rig was common there. Western Crude Oil Co. of Mt. Pleasant operated this rig, which was powered by a 425-horsepower Caterpillar diesel engine and nicknamed "Four and a Quarter Cats." The K-100 Kenworth had no problem doing its job. *Neil Sherff*

The International Transtar 4370 is the single-axle version. Its standard engine was the Cummins NTC 290-horsepower diesel with the standard Fuller RT-910 10-speed transmission. It had a 117-inch BBC. Cummins and Detroit diesels were the engine selections with horsepower from 238 to 434. The wheelbases were 152, 162, and 174 inches. Fuller and Spicer transmissions with up to 16 speeds were all available. The 4370 had a standard 12,000-pound front axle and the rear axle was 23,000 pounds. *International*

Truck-trailer combinations were popular rigs in the western states. Tanker combinations were one of the popular types, like this 359 Peterbilt for Unocal 76. It had a 119-inch BBC. The standard engine for this model was the Cummins FNTC 300 big cam 1V with a standard Fuller Roadranger nine-speed transmission. The front axle was a Rockwell 12,000-pound rating and the rear was a Rockwell 38,000-pound rating. *Author collection*

Here we see a Brockway model 761 LL. The engine choices were Caterpillar, Cummins, and Detroit diesels with horsepower from 230 to 450. The transmission speeds ranged from 5 to 16. The rear-axle ratings were 34,000 to 65,000 pounds on a 186-inch wheelbase. This 761 LL is set up as a dump truck with an optional grille guard. *Brockway*

Canadian truckers liked the new White Freightliner conventional tractor. Mel's Express Ltd. of Bradford and Toronto, Canada, was a hay and produce hauler. On this run it is operating a Can-Car reefer trailer. Most trucks had the bumper guides mounted on the bumper, but they are mounted on the bottom of the fenders on this tractor. *Harry Patterson*

Right: Here we have another W-900 Kenworth sporting a lot of chrome, a nice paint job, and good looks as it pulls a Timpte reefer down the highway. That makes a good match to the tractor. Golden California Packing Co. of Fresno, California, is proud to have this rig hauling for it. A lot of the western trucking companies and private carriers have always had fancy equipment in their fleets. *Neil Sherff*

Opposite bottom: The Diamond REO Raider was a handsome, husky-looking truck. This 1975 model was owned by Lenny Faus, who was the owner of National Transportation, Inc., a refrigerated hauler out of Omaha, Nebraska. This photo was taken in 1987 at the ATHS Convention in Schaumburg, Illinois. Although the sharp-looking Raider was only 12 years old at the time, it was too young to be classified as an antique, but it certainly was a welcome sight. The NTC 350 engine was standard on tandem-axle models, but with the big 1,600-square-inch radiator it could handle the newer, bigger engines like the 8V-92T Detroit or the Cummins KT-450. All the chrome and extra components, along with the long wheelbase, make this a real eye-catching tractor. *Ron Adams*

Above left: In reading through the sales literature, the Autocar is advertised as having durability, reliability, accessibility, and flexibility. This DC9364 is a heavy-duty tractor that was engineered to handle big jobs. The owner of the rig is unknown, but it is a New Jersey–based outfit. The tractor is pulling a Hyster low-bed trailer that hauls a Caterpillar Dozer from job site to job site. The DC9364 is in the 62,000-pound GVW class. Grille guard and skid guards are a must for radiator and engine protection in construction work. A 975-square-inch radiator was standard for this tractor. Two other radiator choices were 1,300 and 1,440 square inches for larger engines. *Harry Patterson* **Above right:** Another tough Autocar is this DC9364 owned by W. D. Mischler of Berkeley Heights, New Jersey. The name "Marge" is painted below the window, so I assume that this big Autocar is being driven by a woman. Extra chrome gives it a nice appearance. The truck also has a grille guard and a skid guard. *Harry Patterson*

Here's another great-looking Diamond REO Raider. The Cummins NTC-350 was the standard engine for tandem-axle models, but with the 1,600-square-inch radiator, bigger engines like the KT-450 and the 8V-92T were offered. A choice of Fuller and Spicer manual and Allison and Cummins Sundstrand automatic transmissions were available. The dump trailer is from Penn Body Company. *Author collection*

Above: This model CHDT-BC Marmon is from the heart of Pennsylvania Dutch country. Rohrer's Quality Seeds is in Smoketown, Lancaster County, Pennsylvania. Marmons seemed to capture the hearts of several truckers in that area. A nice-looking tractor has the honor of pulling the Great Dane trailer with great lettering. The standard engine was the Cummins NTC-350 diesel with high-mount fan and a Fuller RTO-9513 13-speed transmission. *Harry Patterson* **Opposite bottom:** Model 758 is another 700-series Brockway. It had a 90-inch BBC. The engine was the Cummins NH-230 with a Fuller T905C five speed transmission riding on a 174-inch wheelbase. The front axle had a 12,000-pound capacity, and the rear axle had a 23,000-pound capacity. The truck is owned by the Clinton Oil Co. in Clinton, Pennsylvania, and delivered Mobil heating oil. The body is a Decker tank. *Author collection*

Above left: Another model of the Diamond REO Raider is this axle-back version. Steel was standard on the axle-back and aluminum was optional. The BBC on this one is 119 inches. Cummins, Caterpillar, and Detroit diesel were the engine selections. Reed and Hoppes are the owners of this Diamond REO recovery truck that is on call for 24-hour towing. Dual stacks finish it off for an impressive appearance. *Harry Patterson* **Above right:** The W-9000 Ford was a versatile tractor that could be applied to any kind of hauling, in this case livestock. The rig is owned by Schweitzberger and Sons of Temperly, Iowa. The livestock ride in a Wilson possum-belly livestock trailer. The standard engine was a Cummins NH-230 diesel with a Fuller 10-speed transmission. It has an 82-inch BBC sleeper cab. *Harry Patterson*

The Chevrolet Titan 90 was a tractor that could fit into most types of hauling. In this case it's pulling a Wilson livestock trailer for K. Helbert of Lebanon, Virginia. The engine choices were Cummins, Caterpillar, and Detroit diesels with horsepower ranging from 201 to 434. Spicer and Fuller transmissions offered 5 to 13 speeds. *Dick Copello*

Left: Many garbage haulers kept their trucks very clean. This Autocar is a good example. Sano Carting Co. in Keyport, New Jersey, uses this Autocar for garbage collection and compacts the garbage into this Leach body. A lot of chrome, pinstripes, and the four air horns really dress up this rig. *Harry Patterson* **Below:** The year 1976 was the 200th birthday of our great United States, and everybody got in on the action in one way or another to help celebrate the occasion, including Graf Premium Beverages of Milwaukee, Wisconsin. Its piece of the action was this R-model Mack and Trailmobile trailer. The three golden bulldogs stand proud and tall as they ride down the highway on the truck that wears their name. *Mack Trucks*

Hauling heavy equipment required the right kind of equipment. This American crane needed the aid of the Louisville Ford tractor and Rogers low-bed trailer to get from one jobsite to another. Laramie Crane Rentals, Inc., in Detroit, Michigan, was the renter and hauler. *Author collection*

It's time to hit the trail again for another roundup. The order of the day was load 'em up and off we go. Once loaded you had to keep moving in order not to lose any head. L. A. "Sony" Campbell of El Paso, Texas, took on the load with one of his W-900 Kenworth tractors and Wilson possum-belly livestock trailers. The three fuel tanks were used so that they could drive longer distances without having to stop so often on those long runs. The sooner the livestock was delivered, the less chance there was of losing any of the herd. *Harry Patterson*

This is a side shot of a Brockway model 761. The 121-inch BBC shows nicely along with the 48-inch add-on sleeper box. The model 78 all-steel cab is kept cool courtesy of the air conditioner. The engines offered were Cummins, Caterpillar, and Detroit diesels. The front axle is set back 46 inches. The picture was taken in 1981 at the Union 76 truck stop in Kuhnsville, Pennsylvania. *Ron Adams*

Another truck celebrating the U.S. bicentennial is this Diamond REO cab-over-engine for Union Carbide Corp. in Suffield, Connecticut. The Royale had the 88-inch sleeper cab, and Cummins, Caterpillar, and Detroit diesels were the power sources up to 475 horsepower. The wheelbases went in 6-inch increments from 116 inches on single axles and 206 inches on tandems. *Author collection*

The White Road Boss wasn't as noticeable as other tractors. It had a strong steel frame with a steel cab unless the optional lightweight aluminum was specified. Cummins, Caterpillar, and Detroit diesels were the power source choices up to 435 horsepower with Fuller, Spicer, or Allison automatic transmissions. The driver has the comforts of the optional air conditioner. In comparison, by selecting the lightweight aluminum cab, there was a weight savings of 265 pounds. This tractor looks good as it displays the frameless dump trailer. *Ron Adams collection*

Above: Another popular truck-trailer unit is this Peterbilt model 359. The 127-inch BBC is the long hood. The standard engine was the Cummins FNTC 300 Big Cam IV. Optional horsepower went up to 475. The front axle was a 12,000-pound capacity and the optional axle was 20,000 pounds. The rear axles were 38,000 pounds and the optional axles were 48,000 to 50,000 pounds. The rig is owned by Manke Lumber Co. in Tacoma, Washington. The neatly stacked load of lumber looks like they could be 6x6s from the Willamette Lumber Co. *Author collection* **Left:** In 1976, the Louisville line became longer. The new Ford LTL 9000 had a 154-inch BBC, including the sleeper. Wheelbases ranged from 174 to 246 inches. The standard engine was the Cummins NTC-350 with the Fuller RT-910 transmission. Many items on this model that are optional on others were standard equipment. This was the custom tractor. R. L. Leonard is the owner of the tractor and the Utility reefer trailer. *Harry Patterson*

Left: The Diamond REO cab-over-engine made its last appearance in 1976, the last year trucks named Diamond REO were produced. The Queen of Diamonds was leased to Colonial Refrigerated Transportation's western division based in Ontario, California, and pulled a Timpte reefer trailer. *Harry Patterson* **Below:** One of the biggest trucking companies in the 1970s was McLean Trucking Co. in Winston-Salem, North Carolina. The W-9000 pictured has the 52-inch BBC cab and Trailmobile 45-foot van trailer and is one of more than 10,000 units. The standard engine was the Cummins NH-230 with the Fuller T-9058 five-speed. McLean is authorized to provide direct service to 33 states. This is a great-looking outfit. *McLean Trucking Co.* **Opposite top:** Another guest at America's bicentennial was this Brockway 759LL. The 759 had a 90-inch BBC. The engine choices were Caterpillar, Cummins, and Detroit diesels with 5- to 16-speed transmissions with wheelbases up to 171 inches. The body is a Rex mixer. *Brockway Trucks* **Opposite bottom:** The logging business in the Northwest was real busy. One of the log haulers was W. V. Lowell of Coeur d'Alene, Idaho. The tractor is a Mack R 763 S with a Pierce logging trailer. The gross weight was about 100,000 pounds. The picture was taken on Forest Service Road National Forest at Coeur d'Alene, Idaho. The power in this outfit is a Cummins NTC-335. The 700 series had a selection of 10 different engines with horsepower ranging from 225 to 335. *Author collection*

Brockway
SPIRIT OF
76
MALONEY

MACK
Mack
GVW 80000
14

THOMAS MOTORS
Wilmington DE

FORD
Glen Raven
GLEN RAVEN MILLS INC.

Above: Colorful, sparkling, dazzling, shining, and tough is the description of this model 761 Brockway as it shows off at a truck show. This Huskie is strong enough to take on the job it was set up for. The huge balloon tires and Ravens dump body are typical on trucks that are set up for dump-truck work. The Cummins, Caterpillar, and Detroit diesel engines were available with up to 430 horsepower, with Fuller, Spicer, or Allison transmissions with up to 16 forward speeds. I'm sure that the job it did was as good as its appearance. *Brockway* **Opposite top:** The Diamond REO Raiders were seen in all types of hauling, including car hauling. Custom-made car racks were designed to fit each individual tractor. The wheels were a choice of aluminum, disc, or cast spoke. Fuel tank capacities ranged from 60 to 120 gallons. Suspension systems included Hendrickson, Rockwell, Reyco, and Neway. You could order Diamond REO's own sleeper box or install a custom-built cab. Sadly, 1976 was the last year for Diamond REO trucks for the 1970s. *Author collection* **Opposite bottom:** The Ford W-9000 took its place out on the open road with the rest of the trucks. They moved along by the aid of a Cummins, Detroit, or Caterpillar diesel with 350 or more horsepower with a Fuller or Spicer transmission. Glen Raven Mills, Inc., chose the W-9000 to pull the Fruehauf trailer. *Author collection*

Penn Yan Express, Inc., of Penn Yan, New York, served six eastern states. The company operated a fleet of 240 tractors and 485 trailers. There were 19 terminals that covered the area. One of the rigs was this Brockway model 750 and Great Dane trailer. It had a short 90-inch BBC. The standard wheels were cast spoke. The front-axle rated in at 12,000 pounds with the rear rated in at 34,000 pounds. The short BBCs were good for pulling the longer trailers and still remain within the legal length limit. *Author collection*

The DM 600 is the medium-sized truck that covers the 45,000- to 70,000-pound range GVW. The wheelbases ranged from 176 to 234 inches. It had a 98-inch BBC. The power options were from 180 to 285 horsepower. The transmissions ranged from 14 to 18 speeds. The standard wheels are cast spoke. Claussen Concrete was one of the guests that participated in the celebration of the 200th birthday of the United States. *Author collection*

This new Mack Cruiseliner and Great Dane trailer celebrated the U.S. bicentennial in 1976. There were 28 engine sizes available for this truck, from 235 up to 450 horsepower. The Cruiseliner had a 54-inch day cab, 76-inch short sleeper, and a 90-inch full-size sleeper cab. The transmissions were from 9 to 15 forward speed. C-Line Inc. was located in Warwick, Rhode Island. *Author collection*

When this Brockway came off the production line, it was pristine like all newborn trucks. After years of work, wear and tear shows on a truck. I'm sure this one was no different. But thanks to Rich Fonda, Dave Stowell, Todd Grimm, and Peter Grimm, the Brockway model U-762-TL pictured here has been reborn. This 1976 truck has a 12V-71 Detroit diesel that puts out 475 horsepower with a 13-speed Roadranger. The truck is owned by Peter Grimm and resides in Melrose, New York. A lot of chrome and a real eye-catching color scheme make this reborn Huskie a star at the show. The picture was taken at the 2003 ATHS show in Syracuse, New York. *Don Mackenzie*

The Peterbilt was a popular tractor in almost every segment of trucking. Aztec Butane, Inc., in Phoenix, Arizona, chose a model 352 to pull the unknown brand tanker-trailer. The available BBCs were optional 54, 63, 73, and 86 with 50- and 80-inch configurations for special requirements. This cab looks like it could be a 73 or a special 80-inch cab. The standard engine was a Cummins NHC-250 diesel with a Fuller RTO-915 15-speed transmission. *Rick Manz*

CHAPTER 3

1977–1979

In 1977, Brockway became another bygone name in the history books, and Mack made the decision to pen up the Huskie for good and let the bulldog roam free.

In 1979, the low-tilt cab-over-engine had a little change in the grille work. The Bison was added to the Chevrolet line in 1977 and was available in 108- or 116-inch BBC. There were 19 engine selections from 219 to 412 horsepower. The Titan 90 SS was available with a special extra-large grille. Another member was the Bruin, known as model J90. It had a 92.75-inch BBC. By 1979, all Titan 90s had the large grille.

In 1978, the Ford W series was replaced by the new CLT-9000. This model offered 54-, 64-, 76-, 88-, and 110-inch BBC cabs. The 110-inch BBC was the big super-size cab equal to the Kenworth 108-inch cab, White-Freightliner Vanliner 104-inch cab, and Peterbilt 110-inch cab. The F series also continued through the 1970s.

GMC changed its style in 1978 and came out with the new General. It offered a 108- and 116-inch BBC. Another new addition was the Brigadier 8000 and 9500 series. Both series had the 92.75-inch BBC. The Astro 95 also had a new look, which was known as the Special Series, with a huge grille and custom paint and color

The White Road Boss was the upgraded model of the White 4000. This has a 122-inch BBC, which gives it plenty of room for the bigger engines up to 435 horsepower in either Cummins, Caterpillar, or Detroit diesels with either Fuller, Spicer, or Allison automatic transmissions. It looks like this one is getting ready to go out on the road to tow in another tractor. *Author collection*

The Mack R models were popular tractors and were seen everywhere. Maine is the home state for this rig that's shown pulling a Highway reefer trailer. A sleeper box was added for the driver on those nights away from home. *Author collection*

schemes. Many various engines were offered for all models—V-6 and V-8 gasoline, plus the complete line of Detroit and Cummins diesels with high horsepower. A great selection of transmissions was also offered.

Around 1977 and 1978, Kenworth introduced the new Aerodyne cab. The Vit and Aerodyne cabs, and the sleeper boxes for conventionals in Vit and Aerodyne, came out at the same time. The W-900 had a 116.5-inch BBC with the standard hood but allowed a few more inches for the extended hood, to accommodate the big Detroit 12V-71 diesel engine.

A 1977 Mack Cruiseliner brochure lists a selection of 32 various horsepower to choose from between Mack, Cummins, Caterpillar, and Detroit diesels. To go with the Cruiseliner was the Superliner, which was Mack's competition to the West Coast manufacturers. It had a 1,500-square-inch radiator to accommodate the biggest horsepower engines available. The engines for the Superliner were the same as the Cruiseliner. Through the years Mack has always had a big selection of models to choose from for the right job.

Number 306 was a real hot Chevrolet Titan 90 that was known as the *Apache Devil*. It was leased to Shultz Transit, Inc., of Winona, Minnesota, and pulled a Fruehauf reefer trailer. The same engines that were available in most other trucks were also available in the Titan 90 and the same for the transmissions. *Harry Patterson*

A brand-new custom-built Marmon sits on a lot waiting for its new owner. It's a model CHDT-BC with a 119-inch BBC. The standard engine was the Cummins NTC-350 with a Fuller RTO-9513 13-speed transmission. It has a fiberglass tilt hood, a 12,000-pound front axle, a 38,000-pound rear axle, dual 5-inch vertical stacks, and disc wheels. The wheelbases varied according to the customer's specifications. *Author collection*

Here is a 1977 Brockway model 761 LL owned by Peter Grimm. The specs of this particular truck are not known. The standard and optional equipment for this model is a Cummins, Caterpillar, or Detroit diesel with horsepower ranging from 230 to 430 with a 5- to 16-speed transmission on a 186-inch wheelbase. There is one thing for certain, these Grimm guys have a good eye for color and a great talent for truck restoration. The last year for Brockway truck production was 1977. *Don Mackenzie*

Complete food service and restaurant equipment were delivered by S. J. Casper Co. of Eau Claire, Wisconsin. A Ford 8000 was used to deliver the goods. The standard weight on the front axle was 7,000 pounds and the rear axle was 18,500 pounds. The maximum GVW was 35,000 pounds and the maximum GCW was 60,000 pounds. The standard engine was a V175 V-8 with a five-speed transmission. The cab is a 93.3-inch BBC. Behind the cab is a Johnson reefer body with a Thermo-King refrigeration unit. *Author collection*

Milne Truck Lines, Inc., based in Salt Lake City, Utah, served the states of Arizona, Nevada, California, Utah, Idaho, and Wyoming. The fleet consisted of 140 tractors and 460 trailers. One of those rigs is this model 352 Peterbilt with a set of double trailers—one Pike and one unknown. The tractor has a 63-inch BBC cab to pull double trailers within the legal length limit. *Author collection*

Exactly who owned this truck and what it was used for is unknown. The truck is an International Paystar 5000 with a Lufkin body and Lufkin trailer. The Paystar had a choice of 19 engines with horsepower up to 400 and eight various transmissions. It had a 114-inch BBC. Notice that the tires on the truck are smaller than those on the trailer. *Lufkin Trailers*

Above: P-I-E of Oakland, California, has been among the top five carriers in the United States since the 1930s. It has always kept up with the latest developments in the trucking industry. One of those developments was the use of triple trailers, which are only legal in the western states. A K-100 Kenworth was selected to do the chore of pulling the triplets. P-I-E offered nationwide service with a fleet of more than 900 tractors and more than 5,700 line haul trailers. P-I-E became a coast-to-coast carrier in 1966 after purchasing All States Freight Inc. of Akron, Ohio. P-I-E filed for bankruptcy in the late 1980s. *Author collection* **Left:** Another large carrier was Transcon Lines, Inc., of El Segundo, California. Its fleet had more than 700 tractors and 1,900 trailers. Transcon Lines was also a nationwide carrier after the purchase of Kramer-Consolidated Freight Lines, Inc., of Detroit, Michigan, in 1965. The tractor is a White Road Boss and pulls a Strick trailer. This picture was taken in 1981 at the Bandit truck stop in New Smithville, Pennsylvania. Transcon was another company that went bankrupt. *Ron Adams*

Above left: Construction work requires trucks that can withstand the beating they take working on construction sites. The Autocar was made to handle those kinds of conditions. This DC64 B is owned by A. Dance Jr. & Son in Plainfield, New Jersey. A grille guard, skid guard, and construction fenders were the necessary dress for this type of work, along with oversize tires. The extra chrome adds a personal touch. *Harry Patterson* **Above right:** The R-model Mack was a versatile truck all the way around. This one hauled logs for the logging industry. It looks like the truck is carrying its weight, along with the tri-axle trailer. Today all types of setups and axle combinations are found in all industries. Howard Liskis is the owner. *Author collection*

Lots of chrome and fancy paint jobs were a common sight on many Peterbilt trucks, like the 359 shown here. Peterbilt had its own selection of color and paint schemes, but it would also accommodate customers' preferences. This truck has a nicely balanced color scheme and is owned by Sikora and Wilson in Los Angeles, California. It pulls a Utility reefer trailer. *Harry Patterson*

Introduced by GMC in 1977, the new General was suited for all applications. Two are shown in this photo. These two have a 116-inch BBC. The shorter version was 108 inches. Caterpillar, Cummins, and Detroit diesels were offered in both V-6s and V-8s. The transmissions were 10- and 13 speed and Allison automatics. Add-on box sleepers were also available. GMC offered its own color schemes and designs and finished it off with lots of chrome. Single- and tandem-axle versions were available with wheelbases up to 256 inches. *GMC Trucks*

Above: Another new member of the GMC family is the Brigadier. A 92.75-inch BBC could pull the 45-foot trailers. Four Detroit and three Cummins diesel engines were available. Winn-Dixie Stores of Sarasota, Florida, used a Brigadier to pull a Great Dane trailer for delivering merchandise to its chain of food stores. *Author collection* **Left:** The White Road Boss was a popular truck, especially among freight carriers, who stuck to the standard version due to expense. The one shown here is a beefed-up version with options that most companies didn't need, such as extra chrome, a fancy interior, and a custom paint job. The two things that both freight carriers and independents had in common were great-looking trucks and excellent engineering. *White Motor Co.*

Roadway Express, Inc., of Akron, Ohio, was a big user of the White Road Boss. Roadway had terminals that covered 36 states from coast to coast. Roadway chose the 92-inch BBC so 45-foot trailers could be pulled within the legal length limit. A lot of Roadway tractors had trailer-high stacks, which helped keep front and side of the trailer clean. Notice the hand truck hanging on the front of the tractor. *Author collection*

The Mack Superliner was one of the best looking trucks ever produced. This tractor, owned by Barry Geoff of York, Pennsylvania, is a great example. Chrome and a nice color and paint scheme shows off this super bulldog. The specs about this particular truck are unknown, but the engine choices were Cummins, Caterpillar, Detroit, and Mack diesels. Mack claims the Cummins KT-450 and the Mack 300-plus engines work at peak efficiency with a Fuller 10-speed transmission. The Maxidyne engine with the Maxitorque transmission have also proven to be superior performers. This is all helped along by the big 1,500-square-inch radiator for the higher-horsepower engines. This is a real sharp-looking rig with a classy Timpte reefer trailer. *Dick Copello*

The Western Star of the late 1970s had the same basic appearance as those of the early 1970s. The grille was a flat chrome piece with a straight wraparound bumper. There was a choice of a 106- or 120-inch BBC. More than 306 colors were offered for a custom paint job. Four radiators were offered from 927 to 1,445 square inches. This Canadian rig is owned by Samuel Steel Company Ltd. of Mississauga, Ontario. *Author collection*

Keller Transfer Line, Inc., of Grand Rapids, Michigan, was a refrigerated carrier that served Michigan, Indiana, Illinois, and Ohio. A total of 95 tractors and 60 reefer trailers made up the fleet. One of those rigs was this White Road Boss II and Fruehauf reefer trailer. The standard engine in the RB-2 was a Cummins formula 290 with a nine-speed Fuller transmission. It had a 1,600-square-inch radiator to handle engines up to 430 horsepower. There was a 92-, 107-, or a 116-inch BBC. This tractor is the 92-inch BBC. *Neil Sherff*

Above left: A lot of Autocars were designed as big, tough trucks. Looking at this picture, you can see the muscle by the oversize tires and the construction-type fenders. It was designed to pull the Rogers Bros. detachable low-bed trailer for hauling heavy shovels or bulldozers. A grille guard and skid guard were a must for this type of work. This DC9364 was owned by Mitschele Construction Co. of Livingston, New Jersey. *Harry Patterson* **Above right:** The Chevrolet C65 series had a 97.5-inch BBC. It offered wheelbases anywhere from 125 to 254 inches. The standard engine was a 366 V-8 with a four-speed transmission. There was a 7,000-pound front axle and a 30,000-pound rear axle with a GVW range from 19,200 to 50,000 pounds and a GCW of 60,000 pounds. Long Trucking Co. of Cape Charles, Virginia, owned this model C65 and Trailmobile reefer trailer. *Harry Patterson* **Opposite bottom:** Many Peterbilts, like this 359, were dressed in their Sunday best seven days a week as they traveled down the highways. A lot of chrome and custom paint jobs were common on Peterbilt trucks. They had a selection of color schemes and designs to choose from, or the customer could create a personal design. Whatever the choice was, it turned out to be a piece of beauty. *Neil Sherff*

Above left: This Marmon was kept busy pulling a low-bed trailer for All States Trucking of Oakland, California. This was the heavy-haul division of P-I-E of Oakland, which had a reefer special commodities division in each state. The picture was taken at the Trexler truck stop in Kuhnsville, Pennsylvania. *Ron Adams* **Left:** The nickname "Junkyard Dog" doesn't quite fit this 1978 DM 686 SX, but that's what owner Larry Breaun called it. The power options for this model ranged from 180 to 315 horsepower. Wheelbases were offered as high as 280 inches with a GVW of 103,000 pounds. The DM 600 had a 101-inch BBC. A lot of time and hard work goes into keeping this fancy truck looking good. *Author collection* **Above right:** The LTL 9000 was Ford's tractor for owner-operators. It offered a 118-inch BBC or a 154-inch with a sleeper box. The standard engine was a Cummins NTC-350 with a Fuller RT-910 transmission. The range in wheelbases was from 174 to 246 inches. The company, 4 Refrigerated Service, Inc., of Ft. Smith, Arkansas, had this LTL 9000 pull a Utility reefer trailer. *Harry Patterson* **Opposite bottom:** The Kenworth model W-900 had been around for almost 20 years in 1978. Its appearance has not changed much at all. The one item on this tractor that wasn't around 20 years ago was the 60-inch Vit sleeper box. Kenworth was the first to introduce the 60-inch sleeper box. It didn't take long to catch on with the owner-operators. W. R. Hurst, Inc., of Blanding, Utah, liked it enough to have one on his chromed-up W-900. Trailing after is a Wilson outside-frame flatbed trailer. *Neil Sherff*

THERMO KING
FOR
4
15
REF. SER. INC.

KENWORTH

Ham, bacon, and luncheon meat were the products of the Portland Provision Co. The Oregon chief is proud to boast about the products while riding on the side of the Utility reefer trailer. The power piece was a Freightline FL 6364. The standard engine was a Cummins NTA-400 with a Fuller RT-12510 10-speed transmission. Wheelbases varied according to the customer's requirement. The front axle was a Rockwell rated at 12,000 pounds with a Rockwell rear axle rated at a 38,000-pound capacity. The tractor shown here has a 63-inch BBC. *Author collection*

Above: The medium-duty International S-1900 replaced the Loadstar series. The standard engine was the MV-404 188-horsepower with a five-speed manual transmission. The rear axle was 18,500 pounds, while the front axle was 7,500 pounds. The tandem-axle trucks sported a 100.3-inch BBC. The body was made by Timpte. *International* **Left:** One of the most talked about fleet of trucks running the highways in the 1970s was Monfort of Colorado. It became a legend because of the Kenworth trucks it ran with the big 425-horsepower Caterpillar diesel engines. The trucks passed almost everything they came upon, which is how the passing lane became known as the Monfort Lane. This picture shows the fleet in the later 1970s. White Road Commander IIs made up part of the fleet, along with Kenworth and Freightliner cab-over-engine tractors. The trailer fleet was made up of almost all Timpte reefer trailers. *Harry Patterson* **Opposite bottom:** This GMC drives proudly down the highway displaying its latest introduction–the General. By this time air-foils were coming on strong for fuel consumption and better mileage. The engines available were Cummins, Caterpillar, and Detroit diesels along with Fuller, Spicer, and Allison automatic transmissions. It was available in both single- and tandem-axle version with a choice of the 108- or 116-inch BBC. Trailing behind is a Fruehauf trailer. *General Motors*

Right: The new Western Star cab-over-engine had its work cut out trying to prove itself to the trucking public. It didn't have to do much to convince the owner of this tractor that was leased to the Midwest Emery Freight System's Tank Division in Chicago. It has the 88-inch cab. The standard engine was 290 horsepower with options up to 525 horsepower. Midwest Emery was a major refrigerated carrier that had a tanker and steel division. *Neil Sherff* **Below:** The GMC Brigadier was a great tractor where length limits were concerned. It had a 92.75-inch BBC that allowed 45-foot trailers in 55-foot lengths. As seen in this photo, it also worked well for auto haulers. The short cab allowed more volume for hauling cars. The 6-71 engine was standard matched up with a Fuller RT-910 transmission with 5-, 6-, 7-, 9-, 10-, and 13-speed manual and also 4- and 5-speed Allison automatic transmissions. The standard rear axle was 34,000 pounds and the front axle was 9,000 pounds. The standard radiator was 1,235 square inches with a 1,400-square-inch version for larger and more powerful engines. *Author collection*

Above: The model 352 Peterbilt was a popular cab-over-engine with owner-operators. It had a big selection of options to choose from, including the 110-inch super-size cab, which offered 24 more inches of sleeping space. The standard engine was the custom-rated NHC-250 Cummins diesel with a Fuller RTO-915 15-speed with overdrive transmission. This tractor was leased to CDB, Inc., of Grand Rapids, Michigan, which offered contract carrier service nationwide. *Neil Sherff*

Left: Mack trucks found their way into every field of trucking; the towing business was no exception. Apple Towing felt that the R model would be a good selection for the job. A Weld Built wrecker body was placed on the chassis and it was off to work. Chrome, fender mirrors, fender guides, air shield, bug shield, and a great-looking color and paint scheme make this truck look spectacular. *Harry Patterson*

One of the specialized and heavy-hauler carriers that served the United States was Hill & Hill Truck Lines, Inc., of Houston, Texas. Its kind of hauling required special types of trailers, depending on what was to be hauled. In this case it was a simple flatbed load. Towing this load was a fancied-up W-900 Kenworth with the 60-inch Vit sleeper box. Long-wheelbase tractors were popular in the western states. *Neil Sherff*

Above: The International S-2500 series was the replacement for the Fleetstar series. The F-2575 was a full-tandem tractor. It had a 111.6-inch BBC. The standard engine was a 6-71 Detroit diesel with a Fuller RT-910 10-speed Roadranger transmission. The front axle was a 9,000-pound capacity and the rear axle had a 23,000-pound capacity. Wheelbases were 162, 174, or 180 inches. The optional engines were four Detroits, nine Cummins, and four Caterpillar diesels with up to 360 horsepower. The transmission choices were all Fuller and went up to 13 speeds. A Trailmobile tank trailer is the rear partner. *International* **Opposite bottom:** The GMC series 6500 was the medium-duty tandem-axle model with a 97.5-inch BBC. The wheelbases varied from 149 to 209 inches. The standard engine was a 366 V-8 with a five-speed transmission. A 427 V-8 gasoline and a 4-53T Detroit diesel were also available. Clark, Spicer, Fuller, and New Process were manuals and Allisons were automatics. A. R. Johnson of Palatka, Florida, is the owner hauling the undercover load. *Harry Patterson*

The Freightliner conventional was only in production for about five years by 1978 and was well-received by those who purchased them. This model FLC-12064 dual-drive tractor was leased to Matlack, Inc., of Lansdowne, Pennsylvania, which operated in the lower 48 states after their purchase of P-I-E's tanker division. The fleet numbered more than 5,000 units, including owner-operators. The specs of this tractor are not known but the standard engine of this model was a Cummins Formula 300 and a nine-speed Fuller transmission. The GVW was 43,000 pounds and the GCW was 85,000 pounds. Matlack later became part of the Quality Carriers Tank Lines System. *Harry Patterson*

Above: No matter what their make, it seems that wreckers are always fancy. This example is a model 359 Peterbilt. A lot of chrome and a fancy paint job are all part of a custom service offered by the towing company. *Author collection* **Right:** The Superliner was the greatest looking truck that Mack ever made. A big 1,500-square-inch grille, dual chrome stacks, chrome bumper, polished aluminum wheels, fuel tanks, air tanks, quarter fenders, a 60-inch sleeper box, and a colorful custom paint scheme make for one great-looking Mack truck. What we see here is all that was mentioned above. No specs are known about this tractor, but the power offered for this model was a Cummins KT-450, Mack 300 Plus at 315 horsepower, Mack V-8 at 325 horsepower, 8V-92 Detroit, and Caterpillar 3406. A Fuller 10-speed transmission can be matched with the Cummins KT-450 or the Mack 300 Plus and the Maxitorque five-speed with the Maxidyne engine. *Harry Patterson*

Above left: The White Road Boss was available in three different BBCs: 92, 108, and 122 inches. The one pictured here is a 92-inch BBC with an add-on sleeper box. The truck is pulling for S & T Sales with locations in Tampa, Florida; Fresno, California; and Newark, New Jersey. The picture was taken in 1982 at the Trexler truck stop in Kuhnsville, Pennsylvania. *Ron Adams* **Above right:** Stones Express, Inc., in Ridgefield, New Jersey, celebrated 85 years in the trucking business in 1978. A fleet of 340 tractors and trailers served three terminals in New Jersey, Rhode Island, and Massachusetts. One of the rigs in the fleet is this GMC Astro 95 and Strick trailer. Air deflectors were very common on a lot of company tractors from the time. *Stones Express, Inc.*

Ford revamped its previous W series and ended up coming out with a whole new cab-over-engine tractor. This newborn was the CL-9000. It was available in five cab BBCs: 54, 64, 76, 88, and a super-size 110 inch. The standard engine was the Cummins NTC-290 with a Fuller RT-910 10-speed transmission. The front axle was 12,000 pounds and the rear axles were the Rockwell 34,000-pound version. Cummins, Caterpillar, and Detroit diesels were optional. A six-speed Spicer and a host of Fuller transmissions were optional. This paint design is the Hard Fan design. *Ford Motor Co.*

Above: Kenworth was the second truck maker to have a 108-inch extended cab (pictured here) but the first to have the high-rise aerodyne. It had almost all the fixings of a home away from home. Kenworth also offered the dome windows on the 86-inch cab. Dallas Vance Trucking of Flint, Michigan, owns this tractor. *Neil Sherff* **Opposite-top:** The International S-2600 series was known as the severe service line. The F-2674 had a 48-inch setback 12,000-pound front axle. It had a 112-inch BBC. Wheelbases ranged from 148 to 238 inches. The standard engine was a 228-horsepower 6-71 Detroit diesel with a 10-speed transmission. The optional engines were 5 Detroit diesels, 11 Cummins, and 3 Caterpillars that ranged from 228 to 350 horsepower. All optional transmissions were Fuller and two seven-speed Spicers. The flat body is a Fontaine. *International* **Opposite-bottom:** The next new model to come from GMC was the Astro 95 SS (Special Series). On the 87-inch BBC there was a 495-pound reduction in weight. The standard engine was a 6V-92A Detroit diesel with a Spicer seven-speed direct transmission. The front axle weighed in at 12,000 pounds and the Rockwell rear axle came in at 40,000 pounds. A host of custom paint schemes was available. Colonial Express chose the Astro 95 SS but decided on its own color and paint scheme, along with the Great Dane trailer. All the fixings are here to make a nice-looking tractor. *Harry Patterson*

INTERNATIONAL
Fontaine

GMC
Colonial Express
GMC
GMC

Above: The star of the show on September 1, 1978, was a Mack Cruiseliner and a set of Comet trailers. The company in charge was Garrett Freightlines, Inc., of Pocatello, Idaho. This Cruiseliner had a 54-inch BBC nonsleeper cab. The two-axle models had wheelbases of 116 to 176 inches. This must have been Garrett's first try with Mack trucks. *Garrett Freightlines Inc.* **Right:** Stretch the wheelbase and add a dromedary body and you can pack in a lot of volume. As we can see here, the stretch was on a 352 Peterbilt. Dromedary bodies were handy when hauling light freight. W. J. Donovan was the agent for Mayflower in the special operation systems. Aero Mayflower Transit Co. has been moving household goods across the country since the early twentieth century. This picture was taken in 1994 at the Bandit truck stop at New Smithville, Pennsylvania. *Ron Adams*

Left: The International S series was available in several different working operations. The F-2674 was in the severe-service class as a mixer truck. It had a 112-inch BBC with a 48-inch setback front axle with wheelbases ranging from 148 to 238 inches. A 228-horsepower 6-71 Detroit diesel was the standard engine teamed up with a 10-speed transmission. There were 18 optional engines and 12 optional transmissions to choose from. *International* **Below:** The Western Star cab-over-engine performed all different kinds of jobs equal to its competitors. Here we see one setup as a truck-trailer combination typical of western operators. Triplet Trucking is the owner of this setup. A 74-inch sleeper cab was chosen for the operation. The standard engine was a Cummins 290, but engines up to 525 horsepower were available as an option. It's a good possibility that the covered load could be lumber. *Author collection*

UNION
76

3751
3752
301

This Ford 8000 was busy delivering fuel for Union 76. The standard engine was the 175 V-8 with a five-speed transmission. The front axle was 9,000 pounds and the rear axle was 32,000 pounds. The wheels were standard cast spoke. The cab on this truck is a 105.3 BBC. Wheelbases went up to 236 inches. *Author collection*

Opposite bottom: The International 4300 Transtar is seen here in a single-axle version. For CTI it has a 117-inch BBC. The standard engine was the Cummins Formula 290 with a 10-speed Fuller transmission. The front axle was 12,000 pounds. This single-axle tractor was kept busy pulling a set of double hopper-bottom trailers. *Author collection* **Right:** Langendorf Bread and its fleet of trucks has been around for many years. The White Road Commander II is pulling a set of Trailmobile double trailers. The standard engine has 300 horsepower. By the end of the 1970s, several different air scoops were available to improve fuel mileage. *Author collection*

INTERNATIONAL
TRANSTAR II
TRANSTAR

Opposite top: Hauling heavy equipment was different from regular freight runs. A GMC General was selected to do the honors of hauling this LS-112 FMC Link-Belt crane. Although being loaded, the truck cannot leave until the proper "Long Load" or "Wide Load" signs and flags are in place. *Author collection* **Opposite bottom:** The International Transtar II Eagle is available with 33 new paint schemes along with hand pinstriping. The three main diesel engines are offered with up to 450 horsepower. Transmissions were available with up to 14 speeds. The standard front axle had a 12,000-pound rating with the rear axle at 23,000 pounds. It had an 88-inch BBC with wheelbases from 142 to 184 inches. *International* **Right:** The International Transtar Eagle 4300 carries the same load of standard and optional equipment as its Transtar II cab-over-engine partner. The Sale Engineering Bulletin Booklet also offers 33 colorful new paint schemes. A special feature is the name plaque with the owner's name on the dash. *International* **Below:** The Freightliner FLT-8664 was as handsome a tractor as they come. The standard engine was the Cummins NTC-290 with a Fuller RT-910 10-speed transmission. The front axle had a 12,000-pound capacity with the rear axle at 38,000 pounds. The cab sizes offered were 48, 51, 63, 75, 81, 86, 96, and 104 inches on wheelbases from 114 to 263 inches. Optional GVW was up to 62,000 pounds and the GCW was up to 150,000 pounds. Optional engine horsepower was as high as 525. Freightliner offered at least 24 different paint designs on the cab-over-engine model. The tractor you see here is what a Freightliner is all about. *Harry Patterson*

Above: This 359 Peterbilt was taking a break and parked just right for a nice picture. The standard engine was a Cummins FNTC 300 Big Cam IV, with optional engines up to 475 horsepower. The standard transmission was a Fuller nine-speed Roadranger. You had a choice of a 36- or 63-inch sleeper box. The front axle was a 12,000-pound Rockwell and the rear axles were 38,000-pound Rockwells. The standard BBC was 119.25 inches. Wheelbases were built to the customer's requirements. A colorful paint scheme added the finishing touch. The reefer trailer is a Utility. *Dick Copello*

Below: This 1979 model C 11664 DHB Diamond REO was set up as a straight truck with a tank body. It was kept busy delivering oil for the Crosson Oil Co. in Fall River, Massachusetts. It had a 116-inch BBC. The wheelbases ranged from 158 to 242 inches. Engine availability was Cummins, Caterpillar, or Detroit up to 475 horsepower with Fuller and Spicer manuals or Allison automatic with up to 16 speeds. The model shown here has a 48-inch setback front axle. *Author collection*

Above: The Monfort Packing Co. always had colorful trucks running out of Greeley, Colorado. The Kenworth W-900 had the big 60-inch Vit sleeper box. The wheelbase is not known, but it is a mighty nice stretch. A lot of chrome and trailer high stacks makes for a sharp-looking tractor. The reefer trailer is a Great Dane. *Author collection* **Below:** The Ford CL-9000 stands tall with its 110-inch double-bunk cab. The big sleeper offers a 54x80-inch mattress. There were three special dress-up packages: A was an optional interior package, B was the exterior color and paint scheme, and C was exterior chrome items. All this was taken down the road by a standard Cummins NTC-290 diesel engine. Freant Motor Lines of Macon, Georgia, chose its own paint design. *Harry Patterson*

Above: The new Freightliner conventional, by now only six years old, was taking owner-operators by storm. This model FLC 12064 featured a GVW of 43,000 pounds and a GCW of 85,000 pounds. The engine was a standard Cummins NTC-290 diesel with a Fuller RTO-910 10-speed transmission. The tractor and East dump trailer were leased to Tajons, Inc., of Mercer, Pennsylvania. *Author collection* **Below:** Willig Freight Lines, Inc., of San Francisco, California, has been around for many years operating as an interstate and intrastate carrier. Its fleet totals 138 tractors and 280 trailers. This model FLT 6364T single-drive tractor is pulling a set of Utility trailers. The standard engine on this model was the Cummins NTC-290 with a Fuller RTO-910 10-speed transmission. A Rockwell 12,000-pound capacity was the front axle and a Rockwell 23,000-pound capacity was the rear axle. All this rode on a 116-inch wheelbase. *Author collection*

Andrews THE AIRFLOW FRONT

Above: A company that was a nationwide refrigerated carrier and earned the title "The Elite Fleet" was National Carriers, Inc., of Liberal, Kansas. Looking at this tractor we can understand why it received the title. This model 352 Peterbilt is a perfect example. The fleet was made up of all owner-operators. Because everything is so compacted between the front and rear axles, the wheelbase is probably the shortest available on a tandem-axle tractor. *Harry Patterson* **Left:** Although the 6000-series GMC was available as a tractor, this is a 6000 series as a straight truck. It has the new 92-inch BBC. The standard front axle was 7,000 pounds and the rear axle was 15,000 pounds. The engine was a 350 V-8 with a four-speed direct transmission. Wheelbases ranged from 125 to 218 inches. The body is an Andrews. *Aluminum Truck Bodies*

Above: These truck-trailer combinations were popular in the western states since the teens and 1920s. Guth Miller Trucking, Inc., based in California, used these combinations. Its choice on this truck was a model FLP-6364 dual drive. A 63-inch BBC cab covered the NTA-400 Cummins diesel engine that was hooked to a Fuller RT 12510 10-speed transmission. The 100-gallon fuel tank was aluminum. The front axle was a 12,000-pound Rockwell and the rear was a 38,000-pound Rockwell. *Author collection* **Below:** Berger Transfer & Storage Inc. of Minneapolis, Minnesota, was a longtime agent for Allied Van Lines of Broadview, Illinois. One of the trucks in the Berger fleet was this GMC Astro SS. The available cabs were a 55-inch nonsleeper and 75- or 87-inch sleepers. The one shown here is the 87-inch version. The engine choices were Cummins, Caterpillar, and Detroit diesels. A dragfoil was available for all Astro SS for better fuel consumption. *Berger Transfer & Storage, Inc.* **Opposite top:** Extended cabs were becoming very popular with owner-operators during this time. Marsh Bros. of Atlanta, Michigan, owned this 110-inch model 352 Peterbilt and leased it to E. T. Vanderark and Sons in Grand Rapids, Michigan. Cummins, Caterpillar, and Detroit diesel engines were available, along with a host of transmissions. The first model 352 was made on March 14, 1959, and up until its discontinuance through 1980, a total of 28,897 were produced. *Neil Sherff* **Opposite bottom:** The Mack Cruiseliner was an all-around great truck for any application. In this case, we catch one horsing around for Bob Hubbard Horse Transportation, Inc., of Walnut, California. The 90-inch sleeper cab is kept cool with air conditioning. A choice of 31 variations of horsepower were available. In this case a few more horsepower were added if you include the load. *Harry Patterson*

39
MARSH
BROS.
GRAND RAPIDS, MICH.

MACK
625
BOB HUBBARD
616

The Freightliner Vanliner by this time has now been around for 11 years. The roomy 104-inch cab was the choice of many owner-operators, including Nick Sonneman of Albany, Oregon. A decent-looking paint job and dual stacks give this cab a great look. The dual breather bonnets indicate that this tractor could have a high-horsepower diesel engine. Nick was kept cool with the comforts of an air conditioner. A Utility reefer hauls the cargo. *Author collection*

Above: The GMC Astro SS was available in three cab sizes: 55, 75, and 87 inches. Here we see the Astro SS with a 55-inch cab. Two Utility trailers are loaded with bottled water for Sparkletts Drinking Water Corp., which has been in business since 1925. *Author collection* **Opposite bottom:** This DS66 Autocar is on trash detail. V. Fodera is an owner who didn't spare any chrome. Autocars were tough-looking trucks with a big radiator and grille guard staring you in the face. *Harry Patterson*

Above: Polished aluminum, chrome, and a nice color and paint scheme is what this model 359 Peterbilt is all about on the exterior. The standard engine was a Cummins NTC-350 Big Cam II with a Fuller Roadranger 13-speed transmission. It had a 119-inch BBC. Caterpillar and Detroit diesels, along with Spicer and Allison automatic transmissions were available options. A big 1,444-square-inch radiator could handle the higher-horsepower engines. The first model 359 was produced on April 21, 1967, and was in production until 1987. Total production was 64,858 units. *Neil Sherff*

In the early 1980s, Ford introduced a new look for the conventional to go with its new CL-9000 cab-over-engine. This one is the LTL-9000. The rig is leased to Keen Transport, Inc., for heavy hauling. *Dick Copello*

Above: What everybody thought was a ridiculous-looking truck turned out to be a big seller for Kenworth. The T-600, nicknamed the Anteater, took a while to catch on but in time they were all over the country. Swift Transportation Co. of Phoenix, Arizona, used the T-600 for pulling doubles in the western states. Swift is now a nationwide carrier. *Rick Manz* **Opposite top:** In 1981, Peterbilt discontinued the 352 and introduced the new 362. The 352 had a split windshield, but the 362 had a one-piece windshield. This one has the 110-inch cab. The engine brands and transmissions were the same as the old 352. Horsepower went as high as 600 with the optional Cummins NTC- and KT-series diesel engines. The one shown here is owned and operated by Ed Ortwine of Owosso, Michigan, and leased to Robinson Cartage of Grand Rapids, Michigan. *Neil Sherff* **Opposite bottom:** From the early 1930s until 1966, Integral sleepers were popular with drivers. Autocar stopped making it in 1966, but it wasn't gone too many years when the White Road Boss returned with it. Circle C Trucking in Michigan had one leased to pull a Fruehauf reefer trailer. This is a typical Michigan tractor with fine pinstriping. There is no shortage of chrome or polished aluminum. *Harry Patterson*

252
Robinson
CARTAGE

WHITE
TRUCKING
978
WHITE

INDEX

Other titles of interest:

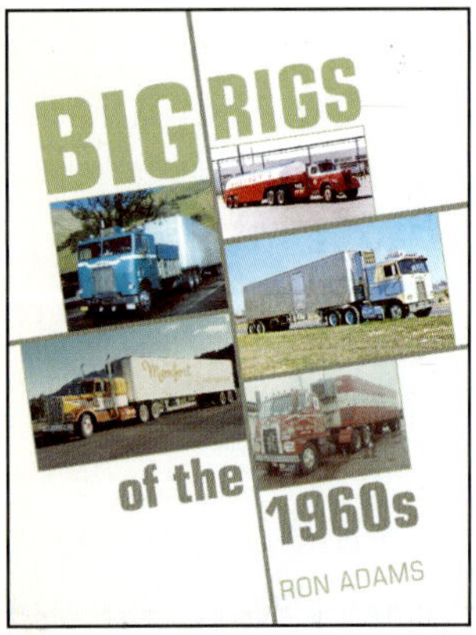

Big Rigs of the 1960s
0-7603-1618-X, 137236

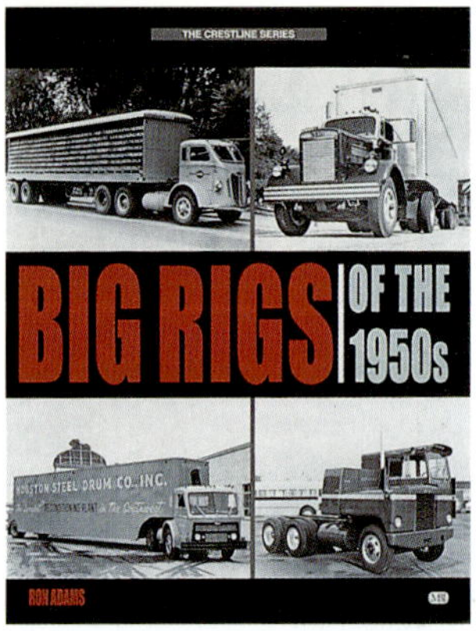

Big Rigs of the 1950s
0-7603-0978-7, 133316

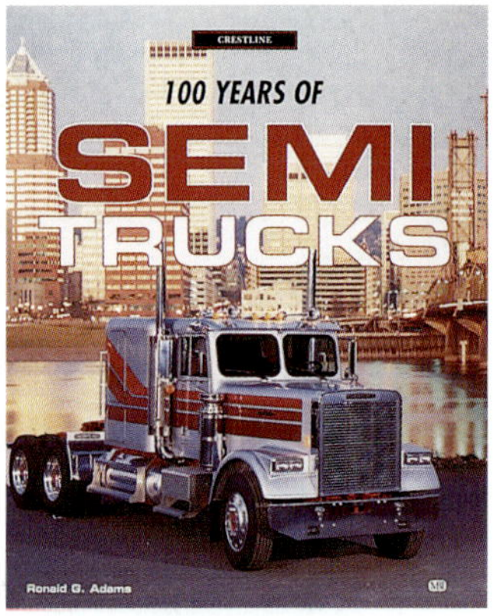

100 Years
of Semi Trucks
0-7603-0769-5, 130098

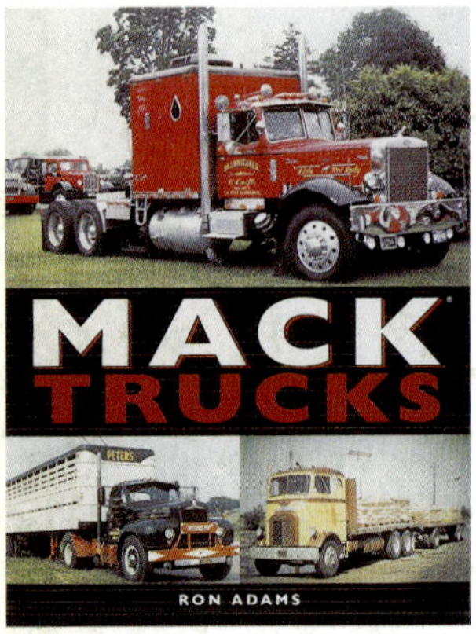

Mack Trucks
0-7603-1237-0, 135127

American Semi Trucks
0-7603-0038-0, 121873

Semi Truck
Color History
0-7603-0351-7, 125097

Peterbilt: The Class
of the Industry
0-7603-1205-2, 133815

Custom Semi
0-7603-2133-7, 139355

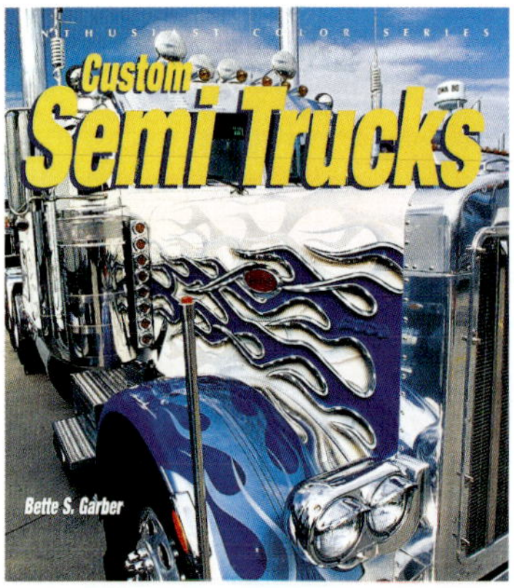

Custom Semi Trucks
0-7603-1459-4, 136257